MATHEMATICS UNRAVELED-
A New Commonsense Approach

dedicated to Ken Sessions
—who liked the idea, but didn't foresee quite all the contingencies...

MATHEMATICS UNRAVELED-
A New Commonsense Approach

By James Kyle

ROBERT E. KRIEGER PUBLISHING COMPANY
MALABAR, FLORIDA

Original Edition 1976
Reprint Edition 1984

Printed and Published by
ROBERT E. KRIEGER PUBLISHING COMPANY, INC.
KRIEGER DRIVE
MALABAR, FLORIDA 32950

Copyright © 1976 by Tab Books
Reprinted by Arrangement

All rights reserved. No part of this book may be reproduced in any form or by any electronic or mechanical means including information storage and retrieval systems without permission in writing from the publisher.

Printed in the United States of America

Library of Congress Cataloging in Publication Data

Kyle, James, writer on electronics.
   Mathematics unraveled.

   Reprint. Originally published: Blue Ridge Summit, Pa.: G/L Tab Books, c1976.
   Bibliography: p.
   Includes index.
   1. Mathematics—1961-   I. Title.
(QA39.2.K94 1984)   510   83-23855
ISBN 0-89874-714-7

Contents

	Preface	7
	Introduction	8
1	**From Counting to Computers**	11
	The Language named Math	
2	**Sets, Et Cetera**	28
	Describing the Indescribable—Order, Relations, and Structure—The Product Set—Mappings	
3	**The Natures of Numbers**	49
	Numbers and Numerals—Eggs and Butter	
4	**Relations, More or Less**	78
	Relations—What are They?—Equality Relations: Equations	
5	**Recipes and Rules**	101
	The Verbs of Arithmetic—More Exotic Ideas—World Without End	
6	**Putting Math to Work**	136
	Describing a Problem—More Complicated Descriptions	
7	**Puzzling it Out**	167
	Truth–Tellers—Checking Up on Things—Who's on First—Number Puzzles	

8   **Some Branches of Math**   **184**
Advanced Numeric Math—Nonnumeric Mathematics

9   **A Whirlwind Look at Calculus**   **209**
Calculi of Calculation—Another Level of Abstraction—Afterword

Appendices

Appendix A   Squares, Cubes, Square Roots, and Cube Roots   242

Appendix B   Common Logarithms   267

Appendix C   Natural Logarithms   269

Appendix D   Values of ϵ^x and ϵ^{-x} 271

Appendix E   Trigonometric Functions   274

Appendix F   Powers of Numbers   275

Index   **277**

Preface

Since first becoming interested in electronics—more years ago than I care to remember now—I have been forcibly reminded (every time I have to calculate a voltage, current, or power level) of the importance of mathematics to today's technology.

When the diffuse designs of destiny led me into the computer industry, it all came back with a slightly different viewpoint.

At first, I rebelled. Math, it seemed, was far too abstract to be really helpful in the real world.

But as time passed I found that the problem was not with mathematics itself, but with the way it had been taught to me. It would be more accurate to say "had not been taught," because when I finally ran into the building-block approach, set forth in the following pages, I found that most of what I "knew" about mathematics was wrong.

Not only did I know little about math at that point, I hadn't even suspected the fun I had been missing. For math, once you understand it, is as much fun as any other form of art can be. In some ways it's like reading a detective story, and in others it's like working a jigsaw puzzle. And in all ways, once you're hooked on it, it's a source of continual fascination.

For many years I had hoped to have an opportunity to present this building-block approach to those outside my own immediate circle of friends and coworkers. The enthusiasm of Ken Sessions for my previous work, *Electronics Unraveled*, has made it possible.

So here it is—and I hope that you find math as much fun as I have, and will.

Jim Kyle

Introduction

Somewhere back in the dim mists which preceded the dawn of history as we know it, one of our distant ancestors discovered the art of counting. Just how or when it happened is not important; the fact that it happened at all may mark the first milepost on the endless path from savagery to civilization.

Ages passed. How many, we cannot know. Man counted. He counted his fellows, his possessions, his enemies, his prey. And along the way he learned some shortcuts to help him keep his tallies.

While man was learning to count his universe, he also learned to use tools. Pressed by the drive to create, to build, he was forced to learn to measure—and he discovered that counting and measurement were much the same sort of thing.

Another milestone had been reached.

More ages passed. Builders and traders learned, and passed on to their successors, the secrets of measurement and of keeping accounts. Some 4000 years ago, the rules were reduced to writing by an Egyptian scribe named Ah-Mose, and his book resides today as the *Rhind* papyrus in the British Museum in London. It is the world's oldest known textbook of mathematics.

Inexorably, man's drive to count and to calculate led him forward, though his progress at first was something less than rapid. As recently as the time of Columbus, arithmetic was unknown to most folk—and considered an idle curiosity by the rest.

Not until the industrial revolution of the 18th century gave birth to the modern age of technology did the importance of math in daily life become apparent.

When it did, education suddenly became a matter of "readin' and 'ritin' and 'rithmetic." An education was not complete unless the ability to do simple sums was imparted.

Even Mark Twain's Huckleberry Finn confessed to knowing his multiplication tables up through "five times seven is forty-two" but, in an echo of an earlier age, doubted that this knowledge would ever prove to be of any value.

From the middle ages on, mathematics expanded in scope. By the time generations of youngsters were being drilled in addition and multiplication, the advanced mathematicians had pushed the frontiers of their knowledge into rarefied heights as far removed from plain arithmetic as arithmetic had been removed from the average person in Columbus' time.

But fast as mathematics grew, once it found fertile ground, the age of technology grew faster. Fields of study that could yield original discoveries and instant fame as recently as the 1930s are now required knowledge for many routine technical courses.

To combat this, educators came up with what was termed the New Math, essentially a rearrangement of the traditional math curriculum. It was an effort to put it all together rather than teaching the subject in piecemeal fashion.

Unfortunately for their noble goal, the New Math must be taught by teachers who themselves learned the math the old way, and who learned to teach the "old" math first. Until the teachers have learned to present "new math" as it was intended to be taught, its effectiveness cannot be judged.

In the meantime, many persons have either been forced by circumstance or have by some accident of interest turned their attention to the subject of mathematics. The fact that you are reading this book indicates that you may be one of these individuals.

Whether your interest is (like my original attack on the subject) forced by necessity, or (as mine grew to become) stimulated by the intellectual challenge itself, you will find here an attempt to present the entire field as one unified entity.

This book will not make a mathematician of you. It may, if successful in its aim, give you a viewpoint from which you can advance through any traditional textbook, taking what you need and ignoring the rest until you need it as well. And if, in so doing, it manages to convince you that all this just might be fun as well as useful, it will have accomplished all that I dared hope for it.

In the early days of human history, wise men did not restrict their wisdom to a single field; one of the early teachers of mathematics was also famed as a philosopher and storyteller. When someone asked him one day how to go about telling a tale, he replied, "Every story has a beginning, a middle, and an end. You should start at the beginning...."

Strangely enough, that's not such a bad way to study math, either; let's start back there with our distant ancestor.

Chapter 1

From Counting to Computers

That distant ancestor who first learned to count may not have qualified for the title *human*, since biologists have proved that many animals have a *sense of number*. Animals that have this "sense" can tell the difference between two objects and three, or even between four and five. Some birds have been known to watch hunters go into a blind and leave it. If two men go in and only one comes back out, the bird stays in hiding until the other man leaves also. If five go in and four come out, the bird's number sense betrays him.

The difference between this action and *counting* is very small. Psychologists explain that to count is to name each number and go through the names in sequence. Since the animal or bird has no name that we know for any number, it cannot "count"—but it is doing the next best thing.

Using this definition of counting, the assignment of names to numbers, the first person to count must have already been capable of speech and communication. In fact, the need to transmit information, such as the number of meat animals in a just-discovered herd or the number of enemies approaching the home caves, may have led to the discovery of speech.

Regardless, we know that counting was one of the first acts of man being made aware of the world about him. The nearest examples to this primitive condition in existence today are isolated tribes in South America and the Pacific islands of New Guinea and the Phillipines.

Some of these primitive tribes have no words for numbers as such, but all of them have words to distinguish the ideas of *one* and *more than one*. One tribe in Bolivia has a word for *alone* and another for *not alone*; others have no words of their own but have borrowed words for *one*, *two*, and *many* from neighboring tribes.

While early man had the ability to count, his counting was limited at first as these examples would indicate.

As culture progressed, so did the art of counting, and in most cultures it was based on the ready-made tally sticks located at the end of each person's two arms. It is no coincidence that in our language the word *digit* has the double meaning of *numeral* and *finger*.

Sociologists studying primitive tribes have written learned papers on the different techniques of finger-counting employed by different tribes. They agree that most primitive persons tally a count of *one* on the little finger of the left hand, then proceed across the left hand, tallying *five* on the left thumb, moving to the right thumb for *six* and on across to the little finger of the right hand for *ten*.

The vote is split down the middle as to whether a count is tallied by closing the extended finger into a fist, or opening a fist into the extended finger.

When I count on my fingers, I habitually use the index finger for *one*, going toward the little finger for *four* and the thumb for *five*, starting with a closed fist and extending fingers as they are counted. Either hand may be used, and the same hand may be used for the second half of the count to ten. Only one primitive tribe was found using this technique.

Studies of kindergarteners indicated that a whole group would use whatever method the first child used, indicating that no really sure pattern existed.

While the fingers formed a convenient set of tally sticks for much early counting (and are used today for the same reason) not all counting was confined to the fingers. Notches were cut into sticks to record counts, and knots tied into string. Pebbles were stacked into piles, and this led to techniques that were forerunners of modern written numeric notation.

Somewhere in this part of the chronicle, *numerals* first made their appearance. The differences between *numbers* and *numerals* are explored in some detail in Chapter 3. For now, we'll simply say that a numeral is a way of naming a number.

Many different types of numerals exist, and others have existed in the past. Our modern numerals, misnamed *Arabic*, include 2 for *two*, 3 for *three*, and so forth. Roman numerals bore more resemblance to finger-counting, with *II* for *two*, *III* for *three*, and *V*—like an opened-out fist—for *five*.

By the time Ah-Mose wrote his textbook, about 1700 B.C., numerals were in common use. Figure 1-1 shows several different types of numerals that were used in early math.

Some of the rules for manipulating numbers had been discovered by the time of Ah-Mose, but it appears that math at that time was more a matter of memorizing facts (like history) than of memorizing rules or learning techniques.

The next milepost in the march of mathematics was furnished by the Greek philosopher Thales, who lived about 600 B.C. Thales studied in Egypt and learned all he could from Egyptian scholars. He went on from that basis to discover the technique of *abstraction*, which permitted him to develop rules rather than having to memorize tables.

While Thales is remembered in the history books, his pupil Pythagoras, who lived from 584 to 495 B.C., is immortalized in plane geometry. The theorem bearing his name is fundamental to most of today's technology—but was known as

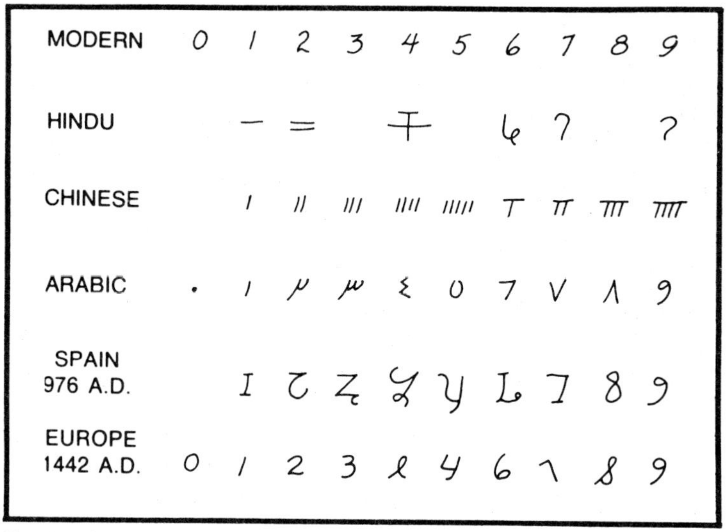

MODERN	0	1	2	3	4	5	6	7	8	9
HINDU		—	=	╪		६	?			?
CHINESE		I	II	III	IIII	IIIII	T	π	πI	πII
ARABIC	·	1	µ	µ	३	0	7	V	ʌ	9
SPAIN 976 A.D.		I	ʗ	ʓ	४	y	⌐	⌐	8	9
EUROPE 1442 A.D.	0	1	2	3	ℓ	Ч	6	⌐	8	9

Fig. 1-1. Several sets of numerals, ranging from earliest known (Hindu) through those used in Europe by 1442 A.D., as compared to modern numerals. Major change in the past 500 years has been in **4** and **5**.

a working rule to the builders of the pyramids. The difference is that Pythagoras proved it as an abstraction, while the builders used it only as a working tool. They built pyramids. Pythagoras built a technological age.

Other, less well known, achievements credited to Pythagoras are the term *mathematics* itself and the division of the field into *discrete* and *continuous* areas for study. The discrete area was further divided into *absolute* and *relative* portions, while the *continuous* area was split into *static* and *dynamic* parts. Pythagoras used other, more colorful names for them, but these are the modern equivalents. The divisions stand to this day, and tend to hide the fact that it is all one field of study.

After Pythagoras, Athens remained the center of the world of mathematcs through the 5th and 4th centuries B.C. With the rise to power of Alexander the Great, the center of civilization shifted to Alexandria, and there Euclid and Archimedes made their major contributions to the field. By this time the Greeks were establishing layers of mysticism atop the crude working tools of the Egyptians; not until the 16th century A.D. would this disguise be removed.

After Alexander's death, the balance of power in the world shifted to the East, and mathematic development during the next thousand years or so flowered in Arabia and India. Before 700 A.D., decimal numbers were in use in India. By 1449 A.D. Al-Kashi, the head of the Samarkand observatory, had used decimal fractions.

With the Crusades, the knowledge moved west again, credited to the Moors and Arabs. By 1500 A.D., math was ready to blossom in Europe. The first English-language text on arithmetic was published in 1522. In France, Francois Vieta introduced the use of *zero* into numbers, and he invented what we know today as *algebra* (though mathematicians qualify the name: *symbolic* algebra).

As the pace of discovery quickened, leading toward the birth of the machine age in 1600, a Scot named John Napier discovered the properties of logarithms (which permit one to perform multiplication by adding), and first made use of the decimal point in calculations. Independently, Edmund Gunter and William Oughtred invented the slide rule, Gunter in 1620 and Oughtred some 12 years later without Gunter's device. Oughtred is credited with introducing the symbol × for multiplication.

In the 16th and early 17th centuries, the roll call of notable mathematical pioneers becomes impossibly long. It includes Rene Descartes, inventor of analytic geometry and of Cartesian coordinates, who lived from 1596 to 1650, and Blaise Pascal, who developed the branch of mathematics known as *probability* (the basis of modern insurance actuarial practice) and who built an adding machine in 1642.

Pierre Fermat left his mark on every branch of mathematics. His "last theorem," unproven to this day, led to the discovery of many fields as his successors sought his "truly wondrous" proof. The marks of Galileo and of Johann Kepler were left not only on mathematics, but on all other sciences as well.

The explosion of invention and discovery culminated with Isaac Newton, born on Christmas, 1642. Best known for his founding of modern physics, Newton is also remembered by mathematicians for his invention of differential calculus—an honor shared with his contemporary, Gottfried Leibniz.

Unlike Newton, Leibniz made most of his contributions solely in the arena of mathematics. In 1673, he built a calculating machine known as the Leibniz wheel, which could add, subtract, multiply, and divide. He is credited as well with the invention of *formal logic*. Both of these achievements are fundamental to today's computers.

Leibniz, a German, was followed by Carl Friedrich Gauss, known as the *prince of mathematics*, who was in his own lifetime acknowledged as the world's greatest mathematician.

Gauss extended number theory to include *complex numbers* (which we will meet again in Chapter 3). Like Fermat, he left his mark on all of mathematics. Many advanced techniques still in use are credited to him. One of the simplest and best known is the *least squares* method for fitting observed data to theoretical predictions, used to adjust the theory to achieve minimum error between prediction and observation.

After Gauss came a deluge of discovery. In the limited space available only the highest points could be touched upon, and those less than adequately. Rather than continue listing bare names and dates, we'll meet the rest as we encounter their discoveries in subsequent chapters. If this taste has stimulated your appetite for mathematical history, dip into James A. Newman's splendid four-volume set, *The World of*

Mathematics; over half of the first volume is devoted to the subject we are allotting half a chapter to. and Newman confesses to being overly selective.

While mathematics was progressing from the practical tool of the Egyptians to its present form. it split into many apparently distinct fields of study.

As we saw. Pythagoras divided it into four major categories but left out the despised *logistica*. the everyday practical arithmetic of builders and buyers. as being beneath the dignity of study.

Already. the sometimes-snobbish distinction between pure and applied mathematics had surfaced. The distinction has led. in this century. to the declaration by a leading mathematician that "...to be pure. mathematics must have no practical use." (a statement with which many nonmathematicians would. unfortunately. heartily agree).

From Galileo to Gauss. the early European mathematicians were primarily philosophers concerned with the real world of physics. They wanted to know why things happened and they used math as a tool. Their mathematical discoveries and inventions were derived largely from practical applications and observations.

As a result. while most of their methods worked. they lacked what pure mathematicians call *rigor*. That is. they could not be logically proven to work properly at all times; nor could they be proven incorrect. Their only "proof" was that they worked.

By the middle of the 19th century. as the study of math became more specialized and so separated from the applied world of the physicists. this lack of rigor began to prove embarrassing. A worldwide project began among mathematiians to restore rigor to all math. similar to the rigor imposed on plane geometry by Euclid centuries before.

The project did not succeed. Instead. it proved that the rigor of Euclid was erroneous (which led. indirectly. to Einstein's theories of relativity). and produced a rigorous proof that no system of mathematics. as such. could be proved rigorously to be consistent. In technical terms. the question of a system's consistency was proved to be an unsolvable problem.

While at first glance this might appear to have been a total waste of time—akin to debating the exact number of angels able to pirouette on the point of a pin—in actual fact it was the opposite. The ancient Greek geometry problem of squaring the circle had occupied the attention of generations of students, until it was proved impossible of solution. The it was ignored and no more effort wasted on it.

Similarly, the search for a consistent system of foundations for mathematics could have occupied generations had it not been proven hopeless at an early stage. The techniques which led to the proof, moreover, turned out to be highly useful in less exotic fields, such as the design of digital computers and telephone switching systems.

In 1854, Engish mathematician George Boole published a slender volume titled *The Laws of Thought*, which has been credited with being "the invention of pure mathematics" and which was, in fact, the link between symbolic logic and numeric computation. Boolean algebra, however, remained very much in the ivory-tower domain for nearly a hundred years.

Another contribution was the invention of *set* theory by Georg Cantor, which was expanded to become *group theory*. Various approaches to symbolic logic and propositional calculus completed the set of techniques available. Just before Godel proved the proof impossible, others published partial proofs.

The debates between the various groups of pure math devotees who took varying positions occupied much of the literature and activity of mathematics through the first third of the present century.

Meanwhile, on the practical-applications front, workers were stumbling over mountains of simple arithmetic. To compute a table of logarithms accurate to six places, for instance, required thousands of hours of tedious arithmeic. To insure accuracy, each computation had to be done many times by many different mathematicians, and the answers compared until agreement was reached.

Since the earliest days of counting, many aids to math have been used. The *abacus* was known in ancient Greece, and is still used in the Orient. The modern counter in any store gets its name from the *counter boards* used for business calculations before paper and pencil became readily available. All these, however, are merely aids to human mental effort. Man searched for mechanical ways to compute.

Napier used a domino-like system of ivory markers which became known as *Napier's bones* to speed his computations. Pascal built his adding machine to help his father in business. Liebniz put together his wheels to simplify astronomical calculations.

Charles Babbage, an eccentric British genius, proposed a *difference engine* to perform computation mechanically, but it was never completed. Two Swedes, Pehr Georg Scheutz and his son Edvard, inspired by Babbage's ideas, built similar machines in 1834 and 1853. The 1853 machine was sold to an observatory in Albany, N.Y., where it remained until 1924.

Babbage's machine, though uncompleted, is often cited as being the direct ancestor of today's computers; the Scheutz machines, which worked, are little known.

Both took their information from a punched card. The idea of using a punched card to convey information was borrowed from the automatic loom invented in 1805 by Joseph Marie Jacquard, which automated the textile industry with over 11,000 of the looms going into use within seven years after their invention.

In 1860, four years before Boole's death and seven years after completion of the second Scheutz machine, an American inventor whose accomplishments were destined to overturn the world's social structure was born. Despite the impact of his ideas, few except specialists know his name today.

Herman Holerith was a tabulator, trained for tedious and exacting work of collecting numbers into tables, when at the age of 20 he became involved with the 1880 census. While the returns of this census were being tabulated, John Billings (in charge of the vital statistics) chanced to remark to Hollerith that "there ought to be some mechanical way of doing this job," mentioning the Jacquard loom.

Hollerith took it from there and implemented the idea. He devised a card in which 288 holes could be punched at specified locations. He made the card the same size as the dollar bill of the time, in order to use available bank equipment for sorting and stacking them. And he devised a way to encode information into a pattern of holes in the cards.

By 1889 he had the patents on his system of tabulating cards, and it was used for the 1890 census. The tabulated results were published only a month after the returns came in, a miracle of speed which astounded statisticians.

Hollerith then left government service, and in 1896 established the Tabulating Machine Company. Businessmen found bookkeeping simpler with "tab cards," and the company prospered.

In 1911, it became the Computer-Tabulating-Recording Company, and three years later attracted an ace business-machine salesman who had left the National Cash Register organization after running afoul of antitrust laws.

The salesman soon rose to lead the company. By 1924, the company (under the leadership of Thomas J. Watson, Sr.) changed its name to International Business Machines Corporation.

Almost from the beginning, Watson and IBM showed their interest in the academic world as well as in the marketplace. In 1929, Watson helped establish the Columbia University's statistical bureau, which was equipped with his firm's standard machines.

Two years later, a special difference tabulator was designed and built for the Columbia facility. By 1933, the director of the bureau, Wallace Eckert, persuaded Watson that the facilities needed to be enlarged.

The resulting Thomas J. Watson Astronomical Computing Bureau was founded in 1937. The year 1937 was especially significant to applied mathematics, because it marks the point at which the burgeoning art of electronics became able to join applied math.

At Iowa State College, that year, Professor John V. Atanasoff designed and began to build an electronic machine to add, subtract, multiply by 2, and divide by 2, using banks of capacitors for a memory, and employing the binary number system.

Workers at Bell Telephone Laboratories, already involved with the nationwide switching network that comprises the Bell Telephone System, in 1937 began construction of a series of relay-operated calculators using standard telephone components.

At Harvard, in the fall of 1937, Howard Aiken wrote a memo which led directly to construction of the Harvard-IBM *automatic sequence controlled* calculator.

A graduate student named Claude E. Shannon was still doing research for his master's thesis. When published in 1938, the thesis became the landmark paper on the application of Boolean algebra to the design of electrical switching networks.

And in faraway Spain, civil war broke out. It was the dress rehearsal for World War II—and wars have speeded the advance of mathematics ever since the first gunners learned that their projectiles obeyed the mathematical formula of ballistics.

As war clouds darkened, construction of the Harvard-IBM machine began in 1939.

Atanasoff's project ground to a halt—but in 1941 it came to the attention of John W. Mauchly. He spent a week with Atanasoff discussing the idea, and later became a coinventor of the world's first electronic calculator.

In September, 1942, a young ordnance lieutenant (just called to service from a spot as math instructor at the University of Michigan) was assignd to take charge of a computing project being carried out by the University of Pennsylvania for the Ballistics Research Laboratory. Mauchly had, by then, summarized his ideas in a memo which was circulating within the ballistics project. It had gained the interest of J.P. Eckert, Jr., a brilliant engineer on the project.

Lt. Herman H. Goldstine saw his duty as: first, expediting the calculations being done by the project (necessary in order to make our weapons usable), and second, pressing for any advances in computational techniques which could help his first duty.

When Mauchly told him of the possibility of doing computation electronically, and Eckert assured him it was within the realm of possibility, he wasted no time. In less than a year, the idea was transformed from a dream into a research contract at the University's Moore School of Electrical Engineering, and on May 31, 1943, work began on the *electronic numerical integrator and computer*—ENIAC.

ENIAC, which was formally dedicated in February 1946 although it had already solved one problem for the Manhattan Project, was the first electronic computer, but the Harvard machine and the relay devices at Bell Telephone Labs both preceded it as the first programmable computers.

Both the Mark I at Harvard and the general-purpose relay computer at BTL went into service in 1944, and both could be set up by means of a prewired plugboard.

Before ENIAC was completed, the personnel of the Moore project had laid out plans for a successor to be known as EDVAC (electronic discrete variable automatic computer). Dr. John von Neumann (one of the greatest mathematicians of modern times) had become involved with the project, and in 1945 wrote his "First Draft of a Report on the EDVAC." This report, while never formally published, became the blueprint for almost all computers built from that day to this.

With war's end, the project began to splinter. Eckert and Mauchly filed patent applications on their inventions (which included the delay-line memory), and formed a partnership (under the name Electronic Control Company) which built the BINAC machine for Lockheed Aircraft, following the general design for EDVAC.

Shortly afterward, they reorganized into the Eckert-Mauchly Computer Corporation, to build a universal automatic computer for commercial use. Their UNIVAC was the first, and for some time the only, programmable computer available for business. The name is still in the industry, although the firm has since been absorbed into one of the large conglomerate corporations.

Goldstine and von Neumann continued work on the EDVAC at Princeton's Institute of Advanced Studies (the academic home of Albert Einstein as well). The history of the computer from the completion of EDSAC (a British version) and EDVAC to date is more one of technical perfection rather than of basic upheaval, though some of the technical advances have lowered cost to the point that the pocket calculator has become common, and home computers may appear at any time.

From the viewpoint of mathematics, the great significance of the computer is that it is freeing the individual from slavery to arithmetic. Emphasis is instead being concentrated on *how* a problem should be solved. Problems which were previously too complex have now become subject to attack. The new emphasis on proper description of the problem had led to new methods for the attack. This has, in turn, revolutionized other parts of mathematics—but it is more a change of *emphasis* than of basic *goals*.

THE LANGUAGE NAMED MATH

Anthropologists tell us that one of the characteristics distinguishing man from most animals is man's use of language.

By using words to communicate, man can pass knowledge from one generation to another. He can give orders or issue instructions. When one person learns to solve a problem, he can tell his comrades of his discovery, and all can then use it—if his description is adequate.

But no natural language such as English, Spanish, French, Japanese, Russian, or Tagalog is necessarily capable of providing adequate descriptions of many mathematical techniques. If you need an example of this, try to describe the technique of subtracting nineteen from twenty-three, using only everyday English and avoiding all the jargon of math which you learned when you were taught how to perform such a subtraction.

The fact of the matter is that mathematics is a language in itself, separate from whatever natural language in which it may happen to be imbedded.

True, an American speaks of one, two, and three while a German would call those same numbers *ein, zwei*, and *drei* and a Spanish mathematician will deal with *uno, dos,* and *tres*, but when an American, German, or Spaniard writes out the problem "$1 + 2 + 3 = $" on a slate, any of the others will fill in the blank space with a 6 no matter what name he gives the number.

Like any language, mathematics is international in its scope—and like any other language, it is meaningless until one learns the vocabulary and the grammar. Fluency comes with practice.

While the idea of mathematics as a language may strike you as being a bit unusual, it's commonplace in the computer industry where dozens of special-purpose languages have been invented to facilitate programing of the complex machines.

The concept of mathematics as a language did not, however, originate in the computer industry. It goes back at least as far as Leibniz' invention of *formal logic*, which is essentially a study of the form in which statements are made, as distinguished from the meaning or content of the statements.

Formal logic is closely allied, in principle, to the diagraming of sentences, which occupied hours of student time in the old-fashioned approach to English grammar.

The idea of math as a language grew, as the devotees of pure mathematics attempted to provide their temple with a firm logical foundation.

While they couldn't provide rigorous foundations for the entire art, they did manage solid proofs of many parts of the system. and in doing so established a system of grammar more consistent than any attained in a natural language.

The present usefulness of mathematics may be due largely to the fact that the grammar of the language is virtually free of irregularities.

In addition to its internal consistency, the language called mathematics differs in one other important respect from the natural language spoken by almost everyone.

Where natural language attempts to describe the real world around us. and all its nouns and verbs are intended to refer to real objects. the invented language of mathematics has nothing at all to do with the real world in any direct sense. Instead. it describes what goes on in a fictional *ideal* universe, where everything is perfect, there are no exceptions to any rules. and all rules are always obeyed.

The internal consistency of the language is made possible because it describes a consistent universe. But nothing in the real world is perfect; the *ideal* universe described by mathematics cannot exist.

This fact—that math is unreal—may be hard to accept. Perhaps the difference between math and reality can be brought out by contrasting a mathematical statement such as "one plus one equals two" and the real-world statement. "If I have one apple and you have another. together we have two apples."

In the math statement. neither you nor anyone else has ever seen a one all by itself. It is a number. an abstract idea, denoting the sole specific characteristic shared in common among all sets with more than zero yet less than two members; as a concrete object. it does not exist.

The math statement declares. however. that if you take one of these nonexistent abstractions and add to it another of the same kind. the two magically merge into a single, different kind of abstraction named a *two*. Most specifically, it will not remain a pair of *ones*.

The real-world statement, on the other hand, is describing things you can see and smell and taste and feel, rather than nonexistent abstractions.

What's more. if you take both apples and place them side by side, they will remain separate and distinct apples rather than merging into some different thing twice as apple-ish as either alone.

This is, of course, the exact opposite of what happens in the ideal world of mathematics, where the two ones merge into a single two. And that's the difference between math and reality.

Obviously, if no relation existed at all between the nonexistent universe of mathematics and the real universe in which we live, then math would at best be a most arcane art form and could have no practical use.

Since, as we have seen, most of mathematics' discoveries have come about in response to practical real-world needs, there must be some connection between them.

As a general truth, the usefulness of any part of the universe of mathematics depends most intimately upon the correspondence which can be established between it and the real world.

Riemannian geometry, to cite one example, was only an amusing fiction which, though logically true, had no possible application—until Albert Einstein realized that it matched with hitherto-unexplained discrepancies in the laws of physics, and produced his Special Theory of Relativity. With correspondence established, the geometry (which did not change) moved from fiction to nonfiction.

Many matches between math and the real world are less obscure. The mathematical statements involved in counting, for instance, were long ago discovered to correspond rather closely to the real-world events of marking off tallies on fingers.

The correspondence in this case between the real and the ideal is so close that for all practical purposes there is no difference; the language of mathematics appears, here, to deal with the real world.

The distinction between the real and the ideal can rapidly be made obvious if you try to count half a gallon of milk; the math operation of counting simply does not correspond with the real-world action of measuring—although it does apply to tallying the number of measures used.

When we come to addition, it's much the same. Addition is, in fact, a shortcut for counting, and so could be expected to have all the characteristics that counting has.

Addition is counting up. The reverse operation, counting down, is called subtraction, and here is where most of us encounter our first serious departure of mathematical fact from real-world fact.

Nearly every student first meeting subtraction emits the anguished cry "But you can't take five from three!" The teacher then explains about negative numbers.

But the student's reaction is totally true so far as the real world is concerned. Only the introduction of "negative number" ideas, which are true in the realm of mathematics but completely fictional in the real world, gives any meaning to the statement $3 - 5 = -2$.

From this point of initial departure, things move thick and fast. If youhave finished conventional grammar-school arithmetic, you have been taught as truth many "rules" of arithmetic.

These rules *are* true, but only in the nonexistent ideal universe of mathematics. In the real world of everyday life, they make sense only because we find it useful to agree that they do. Until one has been initiated through several years of study, the sense is not there.

For instance, most of us feel that, given enough time and enough paper, we could write down to complete and absolute accuracy any number we could imagine.

The mathematical fact is that the vast majority of numbers cannot ever be specified in numerals to absolute accuracy, but we *can* approximate them as accurately as we are willing to.

The difference between the belief and the fact is the difference between "absolute accuracy" and "as accurately as we are willing to," and that's the difference between the *ideal* and the *real* worlds. Absolute accuracy can exist only in the ideal universe; in the real world, we must settle for something less than absolute precision.

Since the world of mathematics is a perfect ideal, it can in principle be completely predictable. Achieving this has been a major goal of mathematicians through the centuries, but the more they learn, the more appears unknown.

In all the branches of math capable of practical application, however, the goal of total predictability has been satisfactorily approximated.

The capability for complete prediction in the universe of mathematics makes the language much simpler than its natural counterparts, since it doesn't have to be capable of variation to match an unpredictable real world.

You may not agree that mathematics is a simpler language than English (or whatever your native tongue happens to be), but the major reason why it seems more complicated is that we all learn our native languages at an early age, with the strongest possible incentive to learn it well and fast—the need to satisfy our individual desires.

Math, on the other hand, doesn't reach us until somewhat later in life, and with much less incentive to learn it either well or rapidly.

Those parts of math for which we do have adequate incentive, such as counting up our pocket money, or keeping sports scores, we learn readily. What makes math seem hard to so many students is not the part of it that happens to interest them but rather the rest, which does not.

In this volume, we're going to attempt to unravel math and show you how it can all be interesting. In order to do this, we're going to have to develop the language—but we're not going to spend page after page here listing the rules, bits, and pieces of it.

Instead, we'll briefly summarize how the language is put together and used, without going into much detail as to why it is this way. Then in subsequent chapters we will go through the various parts of it in more detail. Before we get to the end of the book, the language should be becoming familiar. As in any language, fluency will come with practice once the basic grammar and vocabulary are yours.

Mathematics was once defined as the *science which makes necessary statements*; another authority has called it a *way of saying more and more about less and less*. For our purposes, let's look at it as a language which *makes statements about relations*. These *relations* involve pairs of *operands*, and may be either true or false. The purpose of pure mathematics is to determine the truth or falsity of the statements which are made.

The ideas of *truth* and *falsity* in the language of mathematics are not as complex as they are in the real world; we'll examine them in detail in Chapter 4 and find out how we can prove, reliably, whether a mathematical statement is true or false.

Applied mathematics, in contrast to pure math, deals only with true statements, and uses them to solve problems of the real world.

Before we can say much that will make any sense about the relations which are the heart of mathematics, we have to know what we mean by an *operand*. When we speak of an operand in math, it's something like the noun or pronoun of a natural language.

For instance, every number is an operand. So are the *answers* in arithmetic, and the *variables* and *unknowns* we meet in algebra. The operand, in math, is what we do things to; it is also what we form relations between (but this may not help much until we see, in Chapter 4, what *relations* are in the language of mathematics).

Operands occur in arithmetic problems as the quantities to which things are done; the symbols which tell us what to do are called *operators*, and the complete package of two operands and one operator expressed by the symbols 2 + 3 is often referred to as an *operation*.

The result of an operation is, itself, an operand, and so can be plugged back into another operation or into a relation. Chapter 5 shows us how operators work. The preceding seven paragraphs summarize all that we need to know, at this point, about the structure of the language called mathematics. As we move along, we'll keep adding to this foundation, so that when we get to the point where we're ready to start solving practical problems, we will have all the tools available.

That's the essence of the mathematical method. A mathematician makes statements, and if he doesn't have all the facts he needs, he simply goes ahead with invented "filler" facts to stop the gaps for a first try.

The patterns which result often give him clues as to what facts he really needs and where to look for them, so that the technique which seems so sloppy at first glance is actually a successful method for invention. One of the reasons that it works is that the language of mathematics is just as well suited to dealing with false statements as with true ones, and provides means for testing both truth and falsity.

This same fact leads to a pitfall for the unwary. An unscrupulous person can juggle false statments through the machinery of the language, and make them appear to be true. This was once put succinctly: "Figures don't lie, but liars can figure."

As we move through the basic structure of mathematics, we'll try to avoid the traps. You'll probably discover them for yourself by the time we reach the end of our exploration—and if you do, you may well find yourself hooked on mathematics.

Chapter 2

Sets, Et Cetera

While mathematics is actually both a language and a manner of thinking, most applications of math in the real world involve some aspect of counting. The concepts and ideas involved with counting therefore offer an especially convenient entry point for any attempt to unravel mathematics and remove some of its mystery.

Mathematicians have grouped these ideas under the name *theory of numbers*, and until the advent of New Math it was a graduate-level course. One of the prerequisites for any student interested in taking up number theory was a knowledge of *set theory*, because the formal theory of numbers is built upon the theory of sets as a foundation.

We're following the same approach. Not because it has always been done that way, but because it seems to be the easiest way to become familiar with the ideas. The designers of New Math thought so, too; while theory of numbers is now taught (although not under that name) in the first five grades of grammar school, *set theory* is taught in kindergarten. And since the kiddies don't know that it was once college-level material, they're not confused.

Actually, there's not much about set theory which should confuse anyone. Sets are the basis of almost everything; set theory is simply a defined set of rules for manipulating the ideas involved in describing sets.

This material may appear, at first glance, to belabor obvious points. The reason is simple: the points *are* obvious, but the specific names we will apply to them may not be.

DESCRIBING THE INDESCRIBABLE

One of the major purposes of set theory is to establish defined meanings for the basic words, terms, and phrases we use in mathematics. We can then use these words, terms, and phrases to describe the ideas to which they refer.

Since the whole utility of mathematics is based upon its capability as a language of precise definition, our definitions must also be precise if they are to be useful.

And here we tend to come a cropper. It should appear obvious that definitions and names are intended to serve similar purposes. Each is meant to identify that to which it refers.

It may not be so obvious that a name for something, while serving as a definition of sorts, is *not* an explanation of it.

The purpose of any definition or name for an idea is simply to draw a fence around the idea, so that everything inside the fence is part of the idea, and everything outside the fence is not.

Simply supplying additional names fails to achieve this purpose. If it did not, all teachers could be replaced by dictionaries. We shall see, as we proceed, that complete definition of any idea is literally impossible.

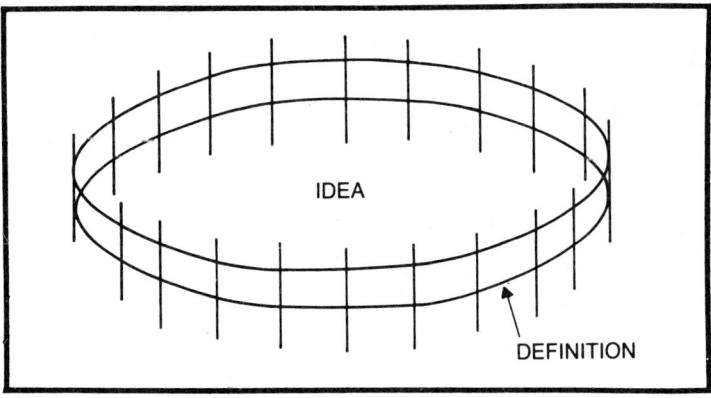

Fig. 2-1. A definition can be thought of as being nothing more than a fence around an idea. Whatever it takes to separate the idea from everything else is adequate.

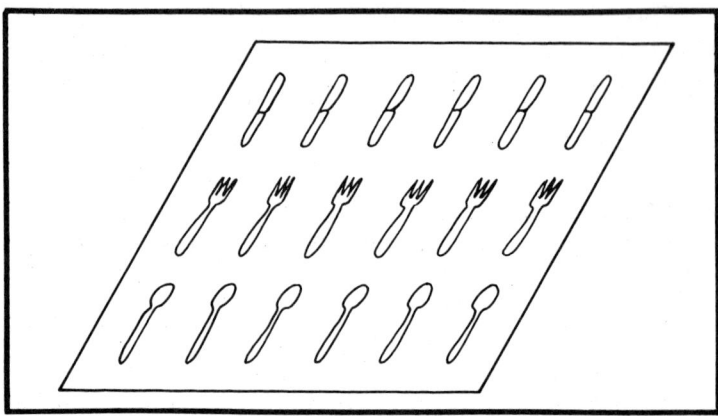

Fig. 2-2. One **set** familiar to almost everyone is a set of silverware. The members of the set are of different shapes and are used for different purposes, but the set is still a single set, separate from everything else.

But while our definitions must be something less than complete and therefore lack absolute precision, this does not imply that they must be so incomplete as to be useless. On the contrary, a definition can always be made more precise by tightening the fence. The necessarily incomplete definitions given here are offered primarily in the form of examples. The various names we use will, for the most part, have no other purpose than to illustrate the ideas.

In each case, the formal name used in the language of mathematics for the idea will be indicated, but we may use any of the alternative names where it will help make the examples clearer

With our ground rules established, let's get on with our excursion into the theory of sets.

The Concept of Sets

A set is a collection, compilation, list, etc. of discrete identifiable objects. The objects may be physical things, intangible ideas, or even other sets (Fig. 2-2). For example, a chess set consists of 16 black pieces, 16 white pieces, and the board. A set of silverware contains many items.

The mathematical meaning of the word *set* is not noticeably different from the everyday meaning of the word; some of the other words involved with set theory are not so lucky.

Each different item in a set is known as a member or an element of the set. Thus. in a chess set. each pawn is an element of the set. although all eight black pawns are superficially alike.

It is possible to define a set which has no members at all. Such a set is called the *empty set*; one example of an empty set is the set of all women who were president of the United States before 1900. Since there were none. the set is empty; it has no members at all.

Sets may be defined in several ways. For small sets, one of the most obvious techniques is to list each and every member of the set. For example. the set of decimal digits may be defined by the list 0. 1. 2. 3. 4. 5. 6. 7. 8. 9. And the chess set may be defined by this list: black pawn. black pawn. black pawn. black pawn. black pawn. black pawn. black pawn. black pawn. black rook. black rook. black knight. black knight. black bishop. black bishop. black queen. black king. white pawn. white pawn. white pawn. white pawn. white pawn. white pawn. white pawn. white pawn. white rook. white rook. white knight. white knight. white bishop. white bishop. white queen. white king. board.

As you can see. the listing technique—called by mathematicians *definition by enumeration*—gets tedious in a hurry when a set has more than just a few members or elements.

Another way to define a set is to name or define a *characteristic* which is common to each and every member of the set. This defining characteristic need not be physical; any characteristic suffices.

Some examples of sets defined by *characteristic* are "the set of all human beings" with the characteristic being *human*; "the set of decimal digits" in which the characeristic is being a decimal digit: and the empty set. in which the characteristic is shared by no members.

Subsets. Any set which has more than one member may be divided into *subsets*. Each subset (Fig. 2-3) is, itself, a set, defined in some manner (whether by listing or by definition). Any specific set may be divided into several sets of subsets unless it has only one member.

Strictly speaking. every set is a subset of itself, and contains the empty set as a subset, in formal set theory. This is necessary in order to achieve consistency in some of the

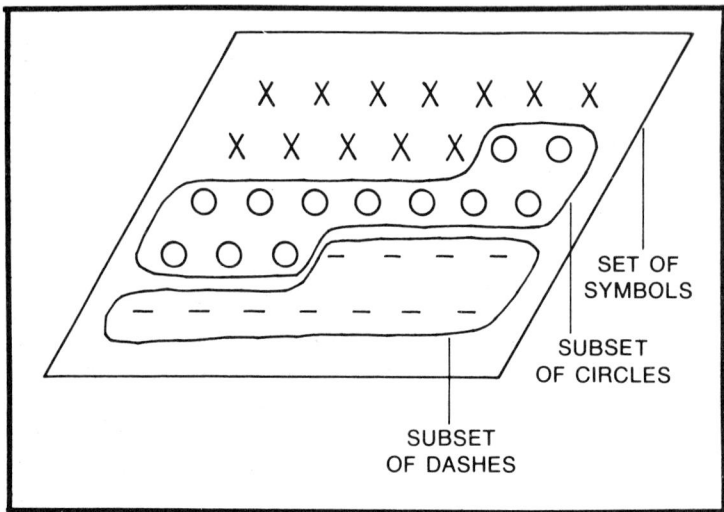

Fig. 2-3. In this set of symbols, two subsets have been partitioned out. One is the subset consisting of all the circles in the set; the other is the subset made up of all the dashes.

higher realms of set theory. In general practice, a subset usually means only a part of the original set; in mathematical language this is called a *proper subset*.

For now, whenever we use the term *subset* by itself we will mean a *proper subset*. When it is necessary to refer to a subset in the strict mathematical sense, we will call attention to it.

Sets with only one member cannot have proper subsets. Thus, the only possible subsets of single-member sets are the set itself and the empty set. Sets having more than one member may be divided into subsets in several ways.

For example, the set of *all human beings* may be divided into the subset of *all male human s* and the subset of *all female humans*. It may also be divided into the subset of *all adults* and the subset of *all children*. Another division might be the subset of *all living humans* and the subset of *all dead humans*.

A specific element of a set may be a member of many subsets of that set. Continuing the example, a specific human may be living, adult, and male.

The whole process of definition or description actually consists of dividing the total universe into appropriate subsets, and selecting those subsets which contain the object being described. The list of those subsets is the description of the object.

We often use the word *classification* for this division process, and in math the word *class* is often used as a synonym for *set*. To classify an object, then, is to assign it membership in some set or sets. The entire process of learning, which begins at birth and continues for life, is one of continual classification.

The list of all sets or classes into which objects or experiences may be classified is equivalent to the total of all knowledge, since learning is equivalent to classification.

But we do not know everything. This implies that objects or experiences exist which may not have any set to match, or at the very least that they have not yet been classified. So far as we can determine, the total universe has no finite limits. Thus, there must be something beyond our present knowledge. And this fact alone means that we cannot divide the universe into a finite number of subsets, unless we cheat by giving one an undefined number of members and labeling it *unknown*. This is in fact what we do.

But since we cannot define the universe in a finite number of completely defined subsets, we can never be certain that any of our supposedly precise definitions do not contain something which should really be in the ill-defined unknown subset. Thus, our definitions will always suffer some lack of accuracy.

We can always increase the accuracy and precision by setting limits of tolerance. This is, in effect, taking the unknown element we know is always present, and assigning it a place. But complete accuracy is simply not possible.

The Universe. In discussion of sets and set theory, the word *universe* has a special meaning. Until now, we have been using it in its everyday sense—"everything there is or can be" is an alternate description of the concept.

But in mathematics, the *universe of discourse* is a set, defined as *the set of all elements being discussed at the moment*. Everything else is outside the universe and does not exist so far as the present discussion is concerned.

During any mathematical discussion, when we define a set to be discussed, it automatically becomes a subset of the universe of discourse.

For example, in the previous section we defined (within the total physical universe or cosmos) the set of all human beings. If we take that set as our universe of discourse, we are discussing only human beings. Within the limits of our discussion, nothing else exists. Our previous subsets—male–female, adult–child, living–dead—are now sets within our universe.

Complements. Any set defined within a universe has a *complement*. The set contains all members of the universe which are described by the set definition; the complement contains all the rest of the universe.

Within the universe of humans, definition of the set *adults* simultaneously defines its complement, *nonadults*. The complement is, of course, itself a set; its defining characteristic is *lack of the defining characteristic of the original set*.

For example, in the universe of humans, when we define the set of all males we simultaneously define its complement, the set of nonmales. The set of all living humans defines its complement, all dead humans.

Note tht the complements in this example are almost—but not precisely—identical to the matching subsets defined earlier. That is, the set of all nonmales is usually considered to be identical to the set of all females, but a few persons have lived who have been neither male nor female. These persons were members of the set of nonmales, but they were not members of the set of all females.

Similarly, most adolescents would claim that they are not members of the set of children, but few adults would grant them membership in the set of adults. Thus, they are nonadults yet they are not children. This furnishes an example of a set divided into more than two subsets, which are mutually exclusive, and at the same time shows how the precision imposed by the strict definitions involved in set theory illuminates the error of many common assumptions.

Supersets and Set Operators

The elements of a set can themselves be sets. This leads to the formation of *sets of sets* or, to use a different name momentarily, lists of lists. The most familiar example of a list of lists is a tabular listing of data, such as the tables of stock market prices in the financial pages of the daily newspapers.

Such tables or, to use their mathematical name, *arrays*, are made up of rows and columns. If you think of each column as being a list or set, then the arrangement of side-by-side columns makes the whole table a list of lists. If, on the other hand, you think of each row as being a separate list or set (such as the opening price, high, low, close, and change for each listed stock in the market reports), then the arrangement of the rows in sequence so that corresponding elements fall in vertical columns makes it a list of lists.

Either way, each column of an array is a subset of the full array; so is each row, but divided according to a different rule. Each specific element or intersection belongs to just one row and one column.

To find a particular entry in a table, we often enter the table from the appropriate row and read from the proper column. Sometimes we reverse the procedure, entering on a column and finding the proper row.

In either case, we find the entry we're looking for at the intersection of column and row. This single entry is a member of the row subset and also of the column subset in which we are interested (Fig. 2-4).

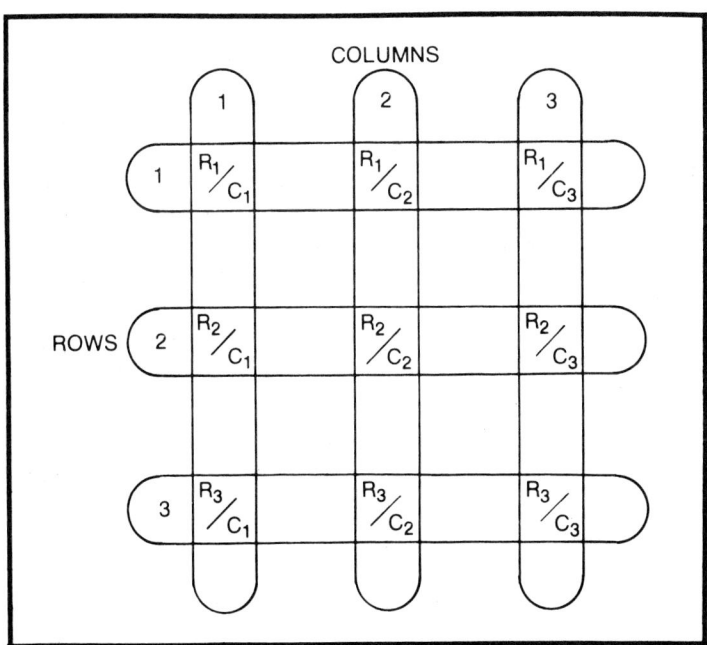

Fig. 2-4. In a table composed of columns and rows, each entry in the table is a member of two different sets at the same time. It's an element in the set of items which makes up its row, and also in the separate set which comprises its column. The whole table, then, is a set of sets, or superset.

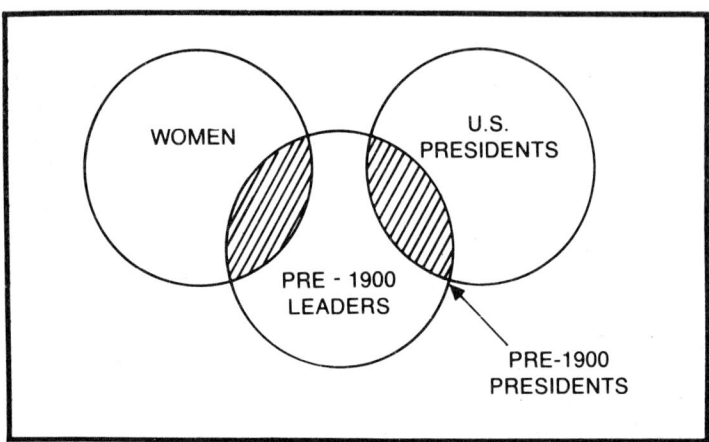

Fig. 2-5. Three nonempty sets may have no mutual intersection, as shown here. Some women were leaders before 1900, so the set of all women and the set of pre-1900 leaders intersect. The United States had presidents before 1900, so the set of U.S. presidents and that of pre-1900 leaders also intersect. But there was no woman president of the U.S. prior to 1900, so the three sets taken together have no mutual intersection.

For instance, if we want to know the high price for General Widget Corporation's stock on yesterday's market, we can take the market listing in the newspaper and search down the *name* column, by row, until we find the "Gen Wt Corp" entry. This tells us that this row reports General Widget's performance. We then look at the *high* column, and have the information we wanted to find.

Note that in such a procedure, we find not one but two intersections. The first is the intersection of the *name* column and a *row* subset for a specific stock; the second is the intersection of the row just found, and the *high* column.

Intersection of Sets. When we deal with sets, we use this same concept of *intersection*; an intersection of two sets or subsets is itself a subset, consisting of all elements which are members of both the original sets.

In the table-entry example, the intersection subset has only one member, which is the entry for which we were looking. An intersection subset can, however, have many members.

For instance, in the universe of humans, the sets of *adult*, *male*, and *living* intersect each other to produce a subset consisting of *all living male adults*. Presumably, this subset has a large number of members.

As we have just seen, intersection is not limited to only two sets or subsets at one time. Any number of sets may intersect. The resulting intersection subset may have one, many, or no members.

In an earlier example, we met the empty set in the form of the set of all women who were president of the United States before 1900. This is actually, as defined, the intersection of at least three sets: the set of women, the set of presidents of the United States, and the set of leaders who held office before 1900. Each of these three sets (Fig. 2-5) has several members; their intersection is empty.

Union of Sets. In the previous section we met the concept of *set intersection*. The intersection of two or more sets is always a subset with fewer members than any of the original sets, unless one of the originals is a subset of another. In that case, the intersection would be identical to the subset.

The other major tool available for working with sets is a concept complementary to that of *intersection*; it's called *union*, and the union of two or more sets is a set which contains as members each and every member of each of the original sets (Fig. 2-6). If we consider for the moment the individual rows of a table to be separate sets, then the union of all these sets is a single set which we may name *table*.

Within this single set, each entry is a separate element and we have no way to relate them to each other by groups. When

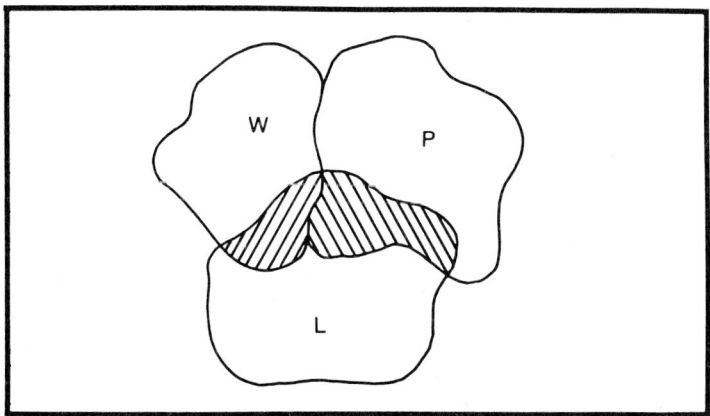

Fig. 2-6. Redrawing the three sets of Fig. 2-5 permits us to show the union of the three sets compactly. Any member of any of the sets is a member of their union, as indicated by the outer boundary line. Internal shading shows the two intersections again.

we consider the table as a set of separate sets. we can locate entries by the intersection method.

But when we want to consult a book filled with different tables. it's convenient to be able to separate one table out by name. That's one of the things the *union* concept makes possible.

Tools of Set Theory. We have. at this point. made the acquaintance of all the conceptual tools of set theory which we will employ. These tools are: the concept of *set* itself. which permits us to manipulate things meaningfully even when we don't know what we're talking about (if only we know what we are not talking about); the *complement* idea. which accounts for all the leftover details; the *universe* limitation. which excludes all the messy details that don't really matter; and the two operators of *intersection* and *union*.   Using these tools. we can define any conceivable idea to whatever degree of precision we may desire. by first defining a universe which we know includes our idea. then defining subsets of this universe to include each aspect of the idea which we can identify.

By continuing this process until all aspects are included as members of defined sets. we establish the limits of the idea. Then. by taking the intersection of all these sets. we pick out the idea itself from all its neighbors and relatives. That intersection is the definition of the idea with which we are dealing.

If. when we take the intersection. we find that it is not sufficiently precise. all we must do to improve the definition is to define additional aspects (additional sets). and repeat the intersection process.

Abstruse as this procedure may appear at first glance. it's the heart of all mathematics. All that set theory does for us is give us tools to trim away all the nonessentials and lay bare the core of any situation.

A sculptor once explained to an onlooker how he was able to produce such perfect woodcarvings: "It's easy." he said. "I just cut away everything that doesn't look like an elephant."

With set theory. we can tell in any situation just what part of our problems to cut away—and if we make a mistake. we can easily backtrack and try a different tack.

Set Diagrams

So far. our discussion of sets and the tools of set theory has been entirely verbal. Since any natural language. English

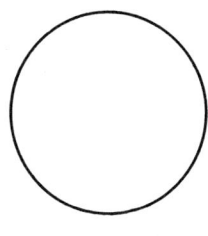

Fig. 2-7. Simplest way to indicate a single set is to draw a circle. This is called a **Venn** diagram, for John Venn, who invented the technique. There's more to come.

included, is itself a set, and the use of words is an intersection process defined only by the user, verbal descriptions of sets and their manipulation are not always adequate.

Fortunately, a simple diagram technique is available to help us. No artistic ability is necessary; only the ability to assign a meaning to a rough sketch is required.

The sketches themselves are known as Venn diagrams after the logician who invented the technique. Figure 2-7 shows the Venn diagram for any single set; it is merely a closed outline which we may think of as the "fence" imposed by the set definition or list. Everything inside the fence is an element of the set; everything outside is not.

Figure 2-8 represents a single set defined from a larger universe. The rectangle represents the universe while the circle represents the set; these symbols are not universal. Some mathematicians draw rectangles for sets, and some draw squiggly outlines like giant jellyfish. The only requirement is that the outline be closed, to provide a complete fence.

In Fig. 2-8, those points outside the circle but inside the rectangle are the complement of the set. Points outside the rectangle do not exist, for the purposes of discussion. When Venn diagrams are used, the rectangle limiting the universe is

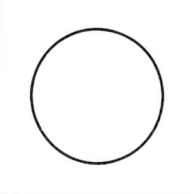

Fig. 2-8. Universe of discourse is indicated in a Venn diagram by putting a box around the circle. The box limits the universe.

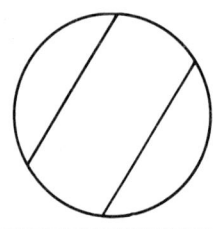

Fig. 2-9. Division of a set into mutually exclusive subsets is shown in a Venn diagram by slicing the set across. Each subset is a different part of the circle.

frequently not drawn since its limits are usually explicitly defined at the beginning of the discussion.

Figure 2-9 shows one way of dividing a set into three subsets using Venn diagrams. You can see that the use of these diagrams, so far, is completely straightforward and no more complicated than idle doodling.

The power of a Venn diagram, though, begins to come to light when two or more sets are involved, as in Fig. 2-10. This is a partial Venn-diagram representation of the sets which we used as examples some pages back to illustrate set intersection.

The sets are A for *adult*, M for *male*, and L for *living* in the universe of humans. Incidentally, this abbreviation of set names is normal practice. Mathematicians insist on writing the set-name abbreviations as bold-face capital letters, but for our purposes this distinction isn't really necessary so long as we remember which letters stand for sets and which for set elements during any discussion.

In Fig. 2-10, notice how the three sets overlap, so that each of the three intersects each of the others independently and at the triple intersection in the center (shown with horizontal shading) which represents the subset of *living male adults*.

The diagram brings out clearly that sets A and M have an intersection which intersects the complement of set L rather than set L itself; this area (vertical shading) represents nonliving male adults and its existence shows clearly that not all adult males are presently living. Notice that this is not the same as the set of dead male adults, since the subset shown with vertical spacing in Fig. 2-10 includes not only the dead, but those who have not yet lived.

Other subsets shown in Fig. 2-10 include all living nonmale adults, all nonliving nonmale adults, all nonliving male nonadults, all living male nonadults, all living nonmale nonadults, and all living or male or adult.

The great power of the diagram technique is that it shows not only the desired intersection or union of sets, but all other relations between the sets being discussed. Draw yourself a copy of Fig. 2-10 and label each of the subsets separately, to prove this for yourself.

There's no magic in using a circle to represent a set. If you have more than three sets to deal with, you can use sausage shapes, rectangles, or what have you. The only absolute requirement is that each set be represented by a closed figure—that is, that the fence around each set be sound, with no gaps in it.

You will probably find, however, that when you try to deal with more than three sets you will need more surfaces than a sheet of paper has in order to draw the diagrams properly. The problem is that there isn't any limit on the number of sets you can talk about in one discussion, but there is a rather low limit

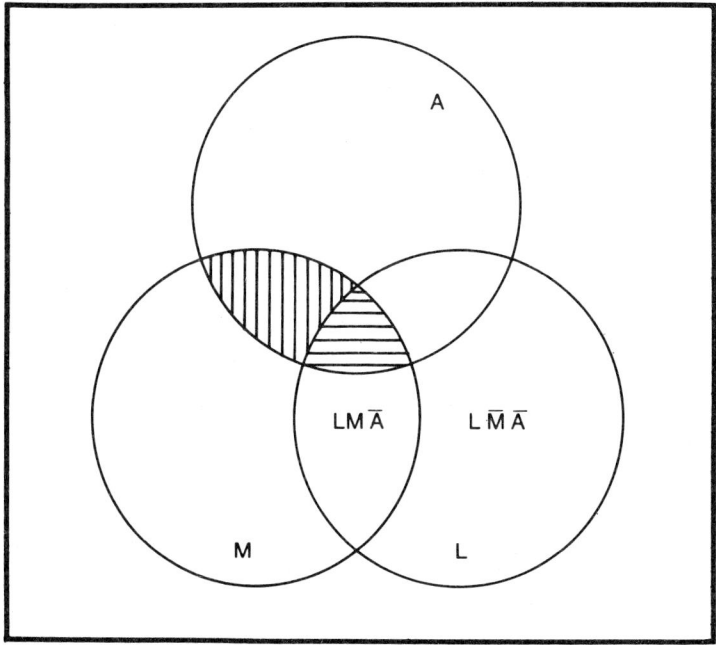

Fig. 2-10. Power of the Venn technique comes to light when it's used to illustrate intersection of the sets of adults, living persons, and males (shown as **A**, **L**, and **M**, respectively). Note that all of the eight possible combinations of three sets are shown in this single diagram. Seven of the combinations are inside one or more of the set circles. The eighth, that of being outside all the sets, is the complement of the union of the three, outside all the circles.

on the number of dimensions available in the real world on which to produce diagrams of them!

The way out of this situation is to work down through the layers of sets, pausing from time to time to redefine the specific subset of current interest as the universe for the time being, and thus eliminating all its competition for diagram space. This trick of continually narrowing your field of vision until the only thing left to see is the problem itself, stripped of all its obscuring detail, is probably the greatest tool which set theory has to offer a would-be problem-solver—and can be applied anywhere, with or without the other tools of set theory.

ORDER, RELATION, AND STRUCTURE

Until now, our discussion of sets has completely left out the idea of order or sequence, but in many sets sequence is one of the most vital properties. In some it is *the* most important—such as the set of all instants in time, if it exists. In this one, sequence is all-important, since without it no member of the set can be distinguished from any other.

Here are some less dramatic examples of sets in which sequence is important: These words are written in symbols chosen from the set of 26 Roman letters (plus a few punctuation marks and case distinctions), but if the symbols are not presented in the proper sequence, no information is transmitted.

Tabular listing of data, likewise, is meaningless unless each of the column subsets (or each of the row subsets) is in a defined sequence, so that the element which forms the intersection with a row subset can be identified.

One of several sets which set theory itself was invented to study was the set of *natural numbers*. This is the group of numbers, not including zero, in which we count. It begins at *1* and continues by integers or whole numbers as far as we need to go. This set, however, has no meaning without sequence. If we did not know that 4 always occurs between 3 and 5, the whole concept of the set would have no point and counting would not be possible.

The importance of order and relation cannot be overemphasized. Definition of this idea is correspondingly difficult; the most essential concepts are usually elusive of definition.

For our purposes at this time, no strict definiton is necessary. When one is required, we will develop it using concepts defined between now and then. For now, we will just think of sequence or *relation* as fixed requirements to the structure of some sets.

When we deal with sets of this type, we shall define the sets by listing their members rather than be means of a defining characteristic, and shall additionally identify them as being *ordered sets*.

When we have defined such an ordered set, any subsequent use we make of it will assume that the order is maintained so that the sequential structure of the set is not disturbed.

For example, if we define the set of decimal digits as merely the numerals 0, 1, 2, 3, 4, 5, 6, 7, 8, 9, without specifying that this is an ordered set, another set defined as the numerals 1, 3, 5, 7, 9, 8, 6, 4, 2, 0 would be equal to it, since each member of the first set is also a member of the second one and all members of each set are thus accounted for.

If, however, we define the first set as an ordered set, then the second set would not be equal to it because no element of the second set is identical to the corresponding element of the first. Actually, only one departure from identity is required to destroy equality when comparing ordered sets.

Once a set is defned as being an ordered set, the sequence of elements is as important as any other characteristic.

A key point here is that the specific order or sequence involved is totally arbitrary, and is chosen when we define the set. Had we defined the second set of digits as the ordered set in this example, then the more familiar first set would have been the one out of sequence within the universe for which our definition held.

Examples of the arbitrariness of our definiion of *order* include simple codes and ciphers, such as those used in newspaper cryptogram puzzles.

To generate a cryptogram, form an array in which the first column is the conventional 26-letter alphabet in its normal sequence (the one we learned in school), and the second column contains the same 26 letters as *another* ordered set, but with different order. The second column, for example, might contain as its elements $n, a, c, i, r, e, m, g, t, o, z, y, x, u, v, w, b, d, f, h, j, k, l, p, q, s$.

To translate any word from its normal spelling into the cryptogram form, replace each letter of the word by the corresponding (same seqence position) letter of the second

43

column. *Computer*, for instance, would become *Cvxwjhrd* in the cryptogram.

Notice that some letters may be the same in both sets; this was done deliberately to emphasize that the choice of sequence has no logic or reason at all when defining an ordered set.

In most cases, though, ordered sets will be attempting to match some observed situation in the real (nonmathematical) world, and the attempt to match will impose some restrictions on the sequence chosen.

THE PRODUCT SET

In our dealings with sets, when we use them to explore the other elements of the language called mathematics, we will make frequent use of tables and arrays, as we just did in illustrating the use of the ordered-set concept for cryptography.

A special kind of table is known as the *product set*, and is used in many ways. The rule which defines the product of two ordered sets as a product set assures us that the result remains ordered. Here's how it works:

Let's assume we have two ordered sets named A and B. The elements and order of set A are defined as u, v, w, x and those of set B as j, k, m.

To form the product set $A*B$, we form an ordered set in which each element contains *two* elements of the original sets. The first part of each of these pairs is an element from set A and the other is from set B. To preserve the order, we take the first element of A and pair it with each element, in turn, of B. When every element of B has been paired with the first element of A, we move to the second element of A and repeat the operation, continuing until all elements of A have been paired with all elements of B.

Thus, the first element of our product is uj, the second is uk, while the fourth is vj. Since set A has 4 elements and set B has 3, the product set must have 4×3 or 12 elements. These are $uj, uk, um, vj, vk, vm, wj, wk, wm, xj, xk, xm$. This product set is an ordered set, preserving the structure of its factors; the elements of set A appear in the same sequence as in set A, and the same is true of the elements of set B.

Had we been forming the product set $B*A$, we would have come up with the same 12 elements but in different sequence: $ju, jv, jw, jx, ku, kv, kw, kx, mu, mv, mw, mx$.

Were sets *A* and *B* ordinary sets rather than ordered sets, these product sets would be considered equal since each consists of the same 12 elements (the reversal of sequence in the element pairs would not be significant in unordered sets).

Since, however, we are dealing with ordered sets, the differences in sequence make the two product sets unequal.

This is a major difference from ordinary arithmetic, where the product of 3×4 is the same as that of 4×3. As we shall see in Chapter 5, arithmetic's multiplication is an application of the rules for unordered sets rather than ordered sets.

Many mathematical operations which yield products as their result behave just like ordered sets, in that the result depends upon which operand is applied first. While the idea may take a bit of getting used to at first, it's actually the most general approach to such combining operations.

Mathematicians describe this sequence-dependent behavior of such operations as *noncommutative*. That is, a commutative operation is not affected by the sequence of operands; ordinary multiplication or addition, for instance, is commutative. When the result *does* depend upon which operand is used first, as in forming a product set, the operation is not commutative.

Sometimes the expression *law of commutativity* is used instead, and a noncommutative operation is said to "not obey the law of commutativity." Mathematicians who say things of this sort are not helping unravel the mysteries of their art; one suspects that they are more interested in impressing their audiences.

MAPPINGS

So far, we have seen how to limit our universe of discourse, define sets (either with or without order as an essential part of their structure), divide them into subsets, take their complement, form their union and intersection, and form their products.

When sets are put to practical use, though, one other essential operation frequently arises. Mathematicians call it *mapping*, but it's actually a matching of one set to another on almost any basis you can imagine.

For instance, let's define the set of all children living on some hypothetical city block as set *C*, and the set of all homes, apartments, and dwelling places on this same block as set *H*.

Since both set C and set H deal with one city block, and the children who form the elements of set C must live in the homes which form the elements of set H, some relationship must exist between the two sets.

This relationship, which associates each child in C with his or her home in H, is what we mean by *mapping*. It can have any of several outcomes:

1. Each member of C matches a different member of H, and every member of H has its corresponding member of C. Every home on the block is the dwelling of a different child. This is called a *one-to-one correspondence* between the sets.
2. Each member of C matches a different member of H, but some members of H have no corresponding members of C. Every child lives in a different home, but some homes have no children. This is a mapping of C *into* H.
3. Some members of C match the same member of H, and every memeber of H has at least one corresponding member of C. Every house has at least one child in residence, but some have more than one. This is called a *many-to-one* correspondence between the sets, and is a mapping of C *onto* H. The distinction between a mapping *into* and a mapping *onto* is that *onto* means that every member of the target set is an *image* (has a corresponding member of the set being mapped); *into* means that some members of the target set may not be images, as in the prior example.
4. Some members of C match the same member of H, but not every member of H has a corresponding member of C. This is, like the second example, a mapping of C into H.
5. No member of C matches any member of H. Either there are no houses or no children on the block. We exclude this possibility by saying that mapping can only be done with nonempty sets.

Another way of looking at the operation of mapping is to consider any mapping of one set into another as the selection of a subset from the product of the two sets. Since the product set contains every possible pairing of the elements of the two sets, any possible mapping must be a subset of the product set.

The feature which distinguishes the mapping subset from all other subsets of the product set is that in a mapping, each element of the set being mapped occurs only once as the first component of an element of the subset (no such restriction applies in the case of the set being mapped into; in our example above, one house could be the home of many children).

Let's look at how this is done using our sets C and H; the members of set C are Jack, Tom, Bob, Jill, Jane, and Nancy; those of set H are red brick, white frame, and pink stucco. We'll abbreviate the house identification as RB, WF, PS to keep the listing of the product set reasonably short.

First, the product set $C*H$ is formed: Jack/RB, Jack/WF, Jack/PS, Tom/RB, Tom/WF, Tom/PS, Bob/RB, Bob/WF, Bob/PS, Jill/RB, Jill/WF, Jill/PS, Jane/RB, Jane/WF, Jane/PS, Nancy/RB, Nancy/WF, Nancy/PS. Since the empty set is a member of every set, we should add " /RB, /WF, /PS, Jack/ , Tom/ , Bob/ , Jill/ , Jane/ , Nancy/ , / " to be complete so far as the product set is concerned, but empty sets and subsets do not take part in the mapping process.

From this product set we could get the following subsets (among others) which would be mappings:

"Jack/RB, Tom/RB, Bob/RB, Jill/RB, Jane/RB, Nancy/RB" if all six children lived in the red brick house.

"Jack/RB, Tom/RB, Bob/WF, Jill/PS, Jane/WF, Nancy/PS" if each house was the home of two children.

"Jack/RB, Tom/WF, Bob/PS, Jill/PS, Jane/PS, Nancy/PS" if four children lived in the pink stucco and each of the other two lived in one of the remaining houses.

In each of these subsets of the product set, each element of set C appears only once, and the rule for formation of a product set assures that one element of set H will appear with it.

In any real case, the actual true mapping would be the one which matched the real-world situation; the others would be equally valid as mappings, but would be false because they would not match the real world.

The two sets involved in a mapping do not have to be ordered, nor do two different sets have to be involved. Our cryptographic example, for instance, which we presented to demonstrate an application of arbitrarily ordered sets, can also be viewed as a mapping of the (unordered) set of 26 letters onto itself so that *a* corresponds to *n*, *b* to *a*, and so forth. The complete mapping would be:

a/*n*, *b*/*a*, *c*/*c*, *d*/*i*, *e*/*r*, *f*/*e*, *g*/*m*, *h*/*g*, *i*/*t*, *j*/*o*, *k*/*z*, *l*/*y*
m/*x*, *n*/*v*, *o*/*v*, *p*/*w*, *q*/*b*, *r*/*d*, *s*/*f*, *t*/*h*, *u*/*j*, *v*/*k*, *w*/*l*, *x*/*p*, *y*/*q*, *z*/*s*.

The similarity between a mapping and a table is becoming more apparent; as we proceed, we will find that almost anything done in mathematics can be described as a *table lookup* process, which means that the ideas involved in the concept of mapping are right at the core of the entire matter.

Chapter 3

The Natures of Numbers

Most common uses of mathematics deal with numbers one way or another. So far, however, we've met only operators, operands, and relations. Where, then, in the language called mathematics do numbers fit? That's what this chapter is all about.

NUMBERS AND NUMERALS

The distinction between *number* and *numeral* is not made simpler by conventional dictionaries, which list them as synonyms one for the other, nor by reference works on mathematics which say only that *number* is a fundamental concept of math which must be strictly distinguished from the concept of *numeral* but give no information as to how to make the distinction.

What is a Number?

The question "What is a number?" may seem ridiculous at first glance—but then, so do many other of the more basic questions dealt with by mathematics. Actually, the idea of number is a rather high level of abstraction, and its definition is correspondingly difficult.

The strictly formal definition for number is probably as easy to comprehend as any other: *number* is defined as that characteristic which several sample sets have in common. For instance, *two* can be defined as the common characteristic shared by the following listed sets: a, b; Tom, Joe; Paris, New York; book, page; one, next.

Each of these sets has more than one member but fewer than three, and the sets have no other characteristic in common. This abstract *two*ness shared by all five sets is what we mean by the number *two*.

In number theory, the concept of numbers is based on the idea of one or *unity*, together with a *successor function* which, when applied to any operand, increases the value or count of that operand by one.

This is a formalization of the physical act of counting or tallying off, and, as such, is not additionally defined. These two primitive postulates are the foundation upon which number theory is built, and no theory can account for its own foundations.

With *one* defined, applying the successor function to it will (by definition) yield a new member of our universe of numbers, which has a value one greater than one.

Since we have already abstracted the idea of *two*ness and named it two by means of several two-member sets, we can map this new entity into any of the two-member sets to demonstrate that the successor of one is identical with what we call two—or, in simpler terms, that one plus one equals two.

If we again apply the successor function, we will obtain the successor of the successor of one, or the successor of two. Again by experiment, we can demonstrate that this is the same as the abstraction we call *three*.

We can continue in this fashion through all the numbers we know, demonstrating the match between each named number and some successor which can trace its ancestry all the way back to our primitive postulates by repeated application of the successor function to unity.

In the earliest days of man's study of mathematics, many persons felt there must be a limit to the number of numbers, or at least to the number of numbers which could be named.

The successor function shows us that no limit to the number of numbers can exist.

Assume, for a moment, that some such limit *did* exist and that we had counted up to it. Applying the successor function to this *limit number* would then yield us a new number larger than the limit.

But by definition, the limit would be a number so large that none could be larger. The contradiction between the assumption that such a number could exist and the results of applying the successor function to it shows that no such limit can exist.

However, for certain purposes, it's still handy to have a name for this idea of the *number which cannot be exceeded*, and so we call it *infinity*. The literal meaning of infinite is "unbounded." Since we have already seen that a concept must be bounded in order to be defined, it's illuminating to consider *infinite* as meaning "undefined," and this gets rid of paradoxes too.

For instance, when we consider infinity as undefined, the problem of the "successor of infinity" (which we just used to prove that no such limit exists) ceases to be a problem. The successor of an undefined quantity must itself be undefined, so the successor of infinity must be infinity itself.

So the answer to our question "What is a number ?" can be either of two answers: *That characteristic common to all sets which can be mapped onto each other in one-to-one correspondence,* or *Some specific successor defined by repeated application of the* successor function *to* unity.

The first of these answers is a real-world concept, based on the idea of physically matching members of various sets. The other is purely mathematical, defning *number* as the result of a not-otherwise-defined operation applied time after time to a not-otherwise-defined quantity.

The fact that both apply to *number* equally well is what makes *number* the major link between computational mathematics and the real world.

Numerals—Names for Numbers

Now that we have the concept of number established, and bearing in mind that number is something essentially different from a numeral, we're ready to decide just what is a numeral.

A numeral is a kind of name for a number, but it's not quite so simple as all that. In the early days people had different names for two persons than they used for two animals, and still other words to mean two rocks or two trees.

Not until the essential idea of *two* as a characteristic of quantity came into existence (completely separate from any other characteristic), could counting become possible.

Some of these special names for numbers still survive. We speak of a pair of socks, for instance, when we mean two. (We also speak of a pair of pants, meaning one, but—like scissors—the one has two legs and takes the plural form of all modifiers.)

Other such special number names include *couple*, for two persons of opposite sex; *brace*, meaning a pair of pistols; *dozen*, for 12 of anything tht can be counted; *gross*, for a dozen dozen or 144; *quire*, for 24 sheets of paper of similar kind, size, and finish; and *ream* for 500 sheets of paper.

None of these special number names are numerals. The numerals in common use today are the symbols for those numbers named (in English) zero, one, two, three, four, five, six, seven, eight, and nine. That is, 0, 1, 2, 3, 4, 5, 6, 7, 8, and 9.

We have also Roman numerals in which I stands for 1, V for 5, X for 10, L for 50, C for 100, D for 500, and M for 1000. Strange as it may seem, addition and subtraction are actually much easier to do in Roman numerals than in the conventional manner, once you get used to it. Because of this, they were commonly used for bookkeeping in European countries until the 18th century, although our conventional numerals were known as early as 1000 A.D. In fact, in 1300 use of our conventional numerals was banned by law for banks and commercial documents, because they were considered easier to forge than the Roman variety.

The Roman numerals are selected letters from the Roman alphabet. Other cultures such as the Greeks and the Hebrews also used letters of their alphabets as numerals, giving rise to the fad of numerology.

When we speak of numbers which are represented in numerals, we sometimes speak of the *digits* of the number. This means the numerals and their positions which are used to write it.

Perhaps the most significant distinction between number and numeral is the fact that any set of numerals is a small and limited set, but the number of possible numbers is not limited.

We can represent any specific number from the unlimited number of possible numbers, using a limited set of numerals, because we have a fixed rule to use in "going around again" which lets us use each numeral as frequently as need be, with a different but defined value at each use.

There's no magic about the set having exactly 10 numerals. either. Computers represent exactly the same numbers. but they do so with a set of only two numerals. In some computer programing systems, a set of 16 numerals is convenient; the letters *A* through *F* are used between 9 and 0 to provide the six extra characters required in the set.

Most authorities believe we use the decimal number system (10 different numerals) simply because we happen to have 10 fingers available for counting. From time to time, other number systems have been proposed. but none successfully.

The only apparent competitor at present to the decimal system is the binary or two-numeral system used in computers. but this is used only inside the machines; outside the mechanical world of the computer, the numbers are converted to other representations to make them more convenient to humans.

The rule which permits us to represent any number with only 10 numerals is called *positional notation* and was developed by Hindu mathematicians before 1000 A.D. It states that in a multidigit number. the rightmost digit shows the number of items counted. the next rightmost indicates the number of times the last numeral has been passed (9 to 0 in the decimal system). and so forth as far to the left as necessary. Every number is assumed to have an unlimited number of zeros extended to its left. so that any digit needed is already there. but these leading zeros are not normally written.

Thus. to count two dozen items in the decimal system, the first nine would be tallied up with single digits using the numerals 1 through 9.

After using 9. the next item would be tallied with two digits; of these. the rightmost would be the first of the set, 0, and the other would be 1 to indicate that the whole set had been gone through one time. giving us 10.

The next nine items would be counted as 11 through 19; the rightmost digit would change with each count, but the other would not.

Upon leaving 9 the second time. the 1 would be counted up to 2 to indicate that the whole set had been used twice. This would give a count of 20: the remaining four of the two dozen items would use tallies 21 through 24.

Since nearly everyone learns this system of using numerals for numbers before starting to school, this explanation may seem unduly complicated. Why not just count?

It's a good question. So long as we stick strictly to the decimal system, we don't really need to know how or why we do what we do. However, mathematics is no limited to the decimal system.

Computers, for instance, use a set of only two numerals but they follow the same rules. This makes the number we know as *4* come up as *100* to the computer, since its count goes *1, 10, 11, 100* rather than *1, 2, 3, 4*.

Each time the set of numerals is exhasuted, the next digit to the left is advanced by one. Since *1* exhausts the set for the computer, *2* must be represented as *10*. We can then show *3* as *11*, but the next count exhausts the numerals at both digit positions, forcing us to three digits and *100* for *4*.

That's why the two-numeral or *binary* system is so hard for humans to follow: it requires too many digits to represent a number. Ten, for instance, becomes *1010*, and one hundred is *1100100*. In the machines, on the other hand, binary is *very* useful because the two numerals can be represented by switches which are on for *1* and off for *0*.

Bases and Base Conversion

We've just been exploring what happens when you have fewer than 10 numerals available with which to represent numbers, and have found that the rules for positional notation permit any number to be represented by any quantity of numerals.

As you might expect, mathematics has special words to describe the situation. A set of numerals establishes a number system, and the number of numerals in the set is known as the *radix* or base of that number system.

Thus, our familiar way of representing numbers with ten different numerals is called the decimal number system and its radix is ten. Sometimes, instead of using the Latin word decimal, we will call it *base-ten*; the two terms mean the same thing.

The computer's way, with only two numerals in its set, is the binary number system; its radix is two.

Any number at all could be used as the base for a number system, but matters would get confusing if a fractional number were used. Generally, only *counting* numbers are used as bases, but use of a negative base has been seriously proposed by some mathematicians (the arithmetic for these, while strictly logical, is a nightmare at first glance).

Besides two and ten, the most common bases are eight (octal number system), twelve (duodecimal system), and sixteen (hexadecimal system). Five is sometimes used in New Math texts to provide practice in converting numbers from one base to another. One of the curious byproducts of the *place value* or positional notation we use, and the rules by which it operates, is that any number greater than one can be written as *10*. While most of us have been trained since childhood to recognize 10 as meaning ten, this is true only if the decimal system is in use.

In mathematics, the representation 10 means only that the base of the number system in use has been reached. In the binary system, this is two; in base-five representation, it is five; in octal, it is eight; in decimal, ten; in duodecimal, twelve; and in hexadecimal, sixteen.

Conceivably, anyone could define a number system with a base of "one million three hundred ninety five thousand four hundred sixty eight" (assuming that enough symbols could be invented to furnish a different one for each number less than this base), and then "10" to that base would be the same number as "1395468" in decimal.

To convert a number from its representation in one base to that in another is not difficult, but can at times be tedious. Essentially what we do is to count up to the number in the system of the starting base, then count back down in that of the new base, but we accomplish this by multiplication or division.

If the new base is smaller than the old, we use repeated division; the remainders at each step provide the digits of the new representation. If the new base is larger than the old, we alternately multiply and add, with the final sum being the new representation.

For instance, to convert *fifteen* from its representation as 15 in decimal to the corresponding 1111 in binary, we will use division. First we divide 15 by 2 to obtain a quotient of 7 and remainder of 1 (we'll abbreviate this result as "7r1" for the rest of these examples, meaning quotient of 7 with remainder 1). This remainder is the rightmost digit of our binary version.

To obtain the next digit, we take our quotient of 7 and divide it by the new base, 2, to get 3r1. The "1" is the second digit of our binary representation.

The third digit is provided when we divide 3 by 2 to get 1r1. Since the quotient is now less than the new base (that is, it is a numeral in the set of the new base as well as the old), it provides the fourth digit and indicates that the conversion is complete.

Let's do it again with a number which doesn't always give the same digits in the new base, such as nine. Remember, we divide by the new base since 2 is less than 10; our first division gives us 4r1, so our first digit is 1.

When we divide the quotient 4 by 2, we get 2r0. In ordinary division, we would ignore the remainder 0, but when we're converting between bases it's important. We use it as the second digit and keep going.

Dividing the quotient 2 by 2 gives 1r0. Again we have a 0 remainder, and the quotient tells us we are through. The converted representation is "1001."

One more time: to convert ten to binary, divide by 2 to get 5r0. Divide 5 by 2 to get 2r1. Divide 2 by 2 to get 1r0. The result is 1010.

Now let's see what happens when we go the other way. Let's convert the binary number *10100* to decimal. Since 10 is larger than 2, we multiply and add this time.

We start with the leftmost digit, and multiply it by the old base (2) to get a result of 2. Add the digit to the right (0); the sum is still 2. Multiply the sum by the old base; the product is *4*.

Add the digit to the right for a sum of *5*. Multiply the sum by 2, giving 10 (decimal). Add the digit to the right; the sum is still 10. Multiply by 2 again for a product of 20. Add the digit to the right; the sum is 20. Since this was the last digit, we stop; the decimal representation of the number is *20*.

Now to prove that all this really works, let's take a number from its decimal representation to base-three, then convert the base-three version to base-eight, and finally change the octal version back to decimal. Let's use the number *1931*.

To convert from base 10 to base 3, we repeatedly divide by 3. We'll abbreviate the steps:

1931/3 = 643r2. 643/3 = 214r1. 214/3 = 71r1. 71/3 = 23r2. 23/3 = 7r2. 7/3 = 2r1. The base-three representation of the number is 2122112.

To convert from base-three to base-eight, we must use the multiply-and-add routine. All the arithmetic for this step must be done in the new base, which now is eight (earlier, we restricted the examples to a new base of 10 to sidestep this problem). The multiplication table for base-eight is rather different from that for the decimal system which we learned in school. Figure 3-1 shows both multiplication tables.

We start by multiplying the leftmost digit, 2, times the old base, 3, giving a product of 6. When we add the next digit, 1, for a sum of 7. Now we multiply 7 times 3 but since we're using octal arithmetic the product is 25 rather than the 21 we would expect from the decimal multiplication table.

To the product of 25 we add the next digit, 2, to get a sum of 27. Now we multiply this, in octal, by 3, to get 105 (when we add the carry of 2 to the product 3×2, we get 10 rather than 8 because 8 is not a numeral in the octal system, but rather the base).

To the 105 we add the next digit, 2, giving 107. Multiplying by 3 again gives 325. Adding the next digit gives 326, and the multiplication yields 1202. Adding the next digit gives 1203, and our multiplication then produces 3611. Adding the final digit, 2, gives us 3613 as the base-eight representation of our number.

If everything has been correct to this point, converting "3613" base-eight back to decimal should return us to 1931 as the representation of the number. Let's see if it does:

Since 10 is larger than 8, we will multiply and add for this conversion. The first digit, 3, times the old base, 8, gives us 24 (the new base is 10, so multiplication is more conventional). To this we add the next digit, 6, for a sum of 30, and multiply the result by 8 to get 240. Adding the next digit gives 241, and multiplying by 8 results in 1928. When we add the final digit, 3, we get 1931, which is what we expected.

The rules for base conversion are simple enough: to convert to a base *smaller* than you're starting from, repeatedly divide and build up the new representation from the remainders. All arithmetic is done in the old base.

To convert to a base which is *larger*, multiply and add. All arithmetic is done in the new base in this case, and the final sum after addition of the rightmost digit is the new representation.

If one of the two bases involved happens to be a power of the other, a simpler technique exists, and this is why the octal

0	0	0	0	0	0	0	0
0	1	2	3	4	5	6	7
0	2	4	6	10	12	14	16
0	3	6	11	14	17	22	㉕
0	4	10	14	20	24	30	34
0	5	12	17	24	31	36	43
0	6	14	22	30	36	44	52
0	7	16	25	34	43	52	61

OCTAL

0	0	0	0	0	0	0	0	0	0
0	1	2	3	4	5	6	7	8	9
0	2	4	6	8	10	12	14	16	18
0	3	6	9	12	15	18	㉑	24	27
0	4	8	12	16	20	24	28	32	36
0	5	10	15	20	25	30	35	40	45
0	6	12	18	24	30	36	42	48	54
0	7	14	21	28	35	42	49	56	63
0	8	16	24	32	40	48	56	64	72
0	9	18	27	36	45	54	63	72	81

DECIMAL

Fig. 3-1. Here are multiplication tables for the octal number system (top) and the more familiar decimal system (bottom). The circled number, 25 in octal and 21 in decimal, is the same quantity, and represents 3×7 in each system. Octal arithmetic is just like ordinary arithmetic, but the numerals have different meanings.

(base 8) and hexadecimal (base 16) systems are popular with computer programers who deal with binary (base 2) representations of numbers.

To convert the binary figure 101001000111 to octal, we simply group the digits by threes from the right (8 is the 3rd power of 2) to get 101 001 000 111, then by using a table convert each group to an octal numeral: 5 1 0 7.

To convert the same binary figure to hexadecimal, we group by fours from the right: 1010 0100 0111. Again we must use a table to convert binary to hexadecimal, finding out that the number is A47.

Those who deal frequently with such conversions usually memorize the tables. That for binary-to-octal is simply mapping; 0/000, 1/001, 2/010, 3/011, 4/100, 5/101, 6/110, 7/111. For hexadecimal, the octal mapping is used with each binary figure preceded by a leading zero, and augmented with 8/1000, 9/1001, A/1010, B/1011, C/1100, D/1101, E/1110, F/1111.

To expand from the octal number 5107 back to binary, we just replace each digit of the octal pattern with the corresponding three binary digits according to the mapping: 101 001 000 111. We then close up the groupings to get 101001000111.

For hexadecimal it works the same way. A47 becomes 1010 0100 0111, which closes up to 101001000111.

The machines do not actually do octal or hexadecimal arithmetic; the base-8 and base-16 number systems are used simply to make it easier for humans to remember and discuss the patterns of binary digits, by converting to something more like everyday numerals.

Positional Notation

Now that we've found out what the New Math texts mean when they speak of bases and converting from one base to another, we can go back for a more exact definition of this thing called *positional notation*, which we've all been using since we learned to write figures.

In school, students are taught to think of the rightmost digit of any written number as the *units* digit. In the decimal or base-ten system, the next digit to the left from the units position is called the *tens* digit.

Each tally of the *tens* digit represents a number just 10 times as great as the numeral itself stands for. Together with the units digit, it gives us the ability to represent any number from 0 to 99 inclusive; this multiplies by 10 the range of our set of numerals. We say that the tens digit has a weight or *significance* of ten.

Its neighbor to the left is the *hundreds* digit, and again the weight or significance of each digit is multiplied tenfold.

As we proceed leftward, each additional digit signifies 10 times the weight of its predecessor. Thus, in any multidigit figure, each numeral stands for a different quantity, and the sum of all these quantities is the number being represented.

If we were using base-eight instead of base-ten, we would speak of the *eights* digit, the *sixty-fours* digit, and so forth. The thing to remember is that *each step to the left would multiply the weight of the digit by the base.*

No matter what the base, this holds true. Positional or place-value notation means simply that the weight assigned to each position in the representation depends upon the base. In the *units* position, the weight is one. In the next position, the weight is equal to the base. The third position carries a significance of the base times the base. The fourth has weight equal to the base times the base, and so forth.

This is, actually, just a restatement of the rule we set forth a couple of sections back, that each time we use up the set of numerals while counting, we increment the next digit to the left and start over. The difference between these rules is the difference between counting and multiplying.

That is, in base-two we have only two numerals: 0 and 1. To count from zero up to five, we can count up 0, 1, 10, (all numerals of set used in the *units* position, so we must take the next one in the *twos* position), 11, 100 (all *units* numerals used, also *twos*, so we must take the next in the *fours* position), and finally 101, for *five*.

On the other hand we can say that we have one *four*, no *two*, and one *unit* in the number, to get $1 \times 4 + 0 \times 2 + 1 \times 1$ for a grand total of 101 (five).

For small numbers, there's no significant difference between the two methods. However, to convert a large number such as 1975 into binary would take nearly two thousand steps by the counting method—a much more tedious procedure than determining that the number is equal to $1 \times 1024 + 1 \times 512 + 1 \times 256 + 1 \times 128 + 0 \times 64 + 1 \times 32 + 1 \times 16 + 0 \times 8 + 1 + 4 + 1 \times 2 + 1 \times 1$, or 11110110111.

In this process, the factors 1024, 512, and so forth are the successive *powers* of the base *two*, and represent the weights or significance of each bit position. Note that the leftmost digit always has the greatest weight, or the most significance. For this reason, it is often called the *most significant digit* or MSD and the rightmost, by contrast, is called the *least significant digit* or LSD.

EGGS AND BUTTER

When we established the concept of "number" as an abstraction of the characteristic of quantity, we tacitly assumed that there is only one kind of "number." This assumption was not true, but it simplified the concept enough to give us a starting point.

Actually, several distinct kinds of "number" exist in mathematics, and this can lead the unwary into confusion.

For instance, the number we call 6 can have at least five distinct meanings or interpretations. It can be considered a *cardinal* number, a *signed integer*, a *rational* number, a *real* number, or a *complex* number.

In each case, both the logical definition and the capability for mathematical manipulation are different. Some operations are not possible with cardinal numbers, but require signed integers. Some may be performed only with rational numbers. Still others apply only to complex numbers.

The differences can best be illustrated by comparing a half-dozen eggs and six pounds of butter (Fig. 3-2). The essential differences between these two uses of the number 6 become apparent if we substitute the fraction one-half for the number six.

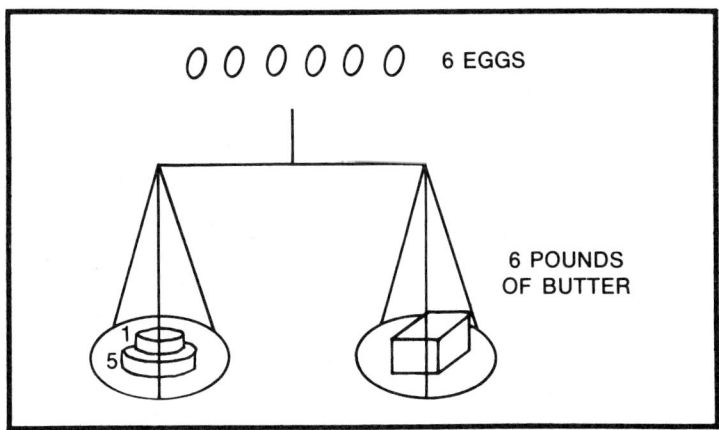

Fig. 3-2. Difference between counting numbers and measurements is shown by six eggs and six pounds of butter. The eggs can be counted; the butter cannot.

61

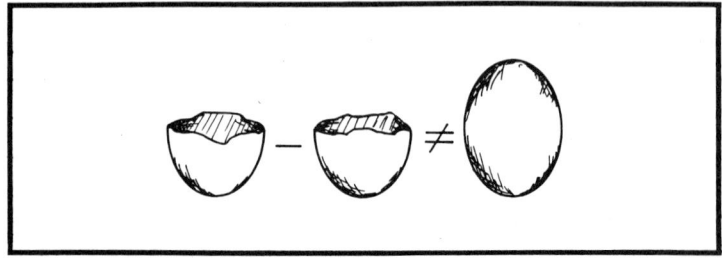

Fig. 3-3. While two half-pound lumps of butter make up a full pound, indistinguishable from the original pound, the same is not true for two half-eggs. There's no way to put them back together to get one unbroken egg.

It's just as easy to measure out a half-pound of butter as it is to measure six pounds, but there's no possible way to count half an egg. If you break it and provide half the shell, half the white, and half the yolk, it's no longer half an egg, but half of what *was* an egg.

That is, you can put two half-pounds of butter together to make a full pound, which cannot be distinguished from any other pound. Two half-eggs obtained by breaking an egg are not the same as one egg (Fig. 3-3), and these is no way to make them so.

The eggs, you see, are counted, using *cardinal* numbers. The butter is measured using *rational* numbers. Division, the operation required to get *one-half*, does not apply to cardinal numbers.

Let's look at these distinctions in more detail, since they show us a thumbnail history of the progress of mathematics and bring us an understanding of why we do some things in arithmetic the way we do.

The Counting Numbers

The counting numbers, or integers, are the ones we used to count the eggs in our example, or to tally on fingers any item that can be counted.

They are also the ones we defined to establish the concept of number, and were historically the first kind of number recognized.

Each cardinal number or cardinal integer is, in fact, defined as *the set of all sets having the same number of elements as a given sample set*. This is almost the same as the definition we learned earlier for the general concept number.

Some authorities call the number *zero* a cardinal integer, and some call it a signed integer. It is defined as *the number of elements in the empty set*. The cardinal integer *one* is then defined as *the number of elements in the* zero *set. Two is the number of elements in the union of the* zero *and* one *sets*.

So long as numbers were used only to count real physical objects, the cardinal integers were able to do all that was necessary, for they can be added and multiplied. They can also be subtracted, so long as the result is still a cardinal integer. That is, you could count 20 cattle going into a pen, and 15 coming out, and subtract 15 from 20 to find there were still 5 cattle in the pen (Fig. 3-4).

However, with cardinal integers it would not be possible to subtract 20 from 15. This would be equivalent to counting 15 cattle going into an empty pen, and 20 coming out.

Fig. 3-4. Subtraction is like counting down. For instance, if 20 cattle enter a pen (top) and later 15 of them leave (bottom), five cattle will still be in the pen. Similarly, 20 minus 15 is 5.

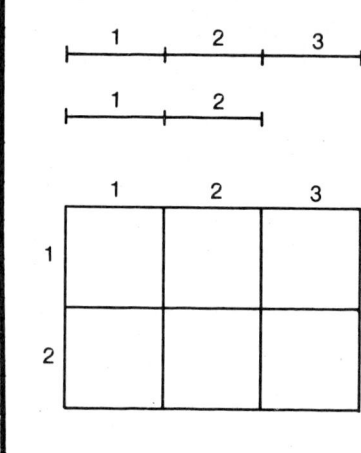

Fig. 3-5. Ratio of three to two is illustrated by different lengths, or **aspect ratio** of rectangle. Greek geometers dealt largely with questions of length, area, and volume, as they developed their rational number system.

When banking was invented and people began keeping books, it became necessary to find a way to represent debts, overdrafts, business losses, and the like. This led to the development of *signed integers*, when the counting process extended to quantities less than zero.

With signed integers, subtracting 20 from 15 is no problem; the answer is −5. This is like having $15 in your checking account and writing a check for $20; you have a $5 overdraft.

The difference between cardinal integers and signed integers is that cardinal numbers extend only from zero (or one, depending on which authority you follow) up to infinity, while signed integers extend from minus infinity through zero to positive infinity.

The negative or *minus* numbers of the signed integers are mirror images of their positive relatives, just as far away from zero but in the opposite direction.

Both sets of integers share the limitation of being counting or *whole* numbers; they have no fractions. The only arithmetic operations generally used with either the set of cardinal integers (usually called *N* or *natural numbers* in math texts) or the set of signed integers (called *I* for *integers*) are addition, multiplication, and subtraction.

As math and the rest of the world progressed, the need to express *ratios* between numbers, such as three to two (Fig. 3-5), became acute. This expression of ratio was written 3/2 or 3:2, and led to the set of rational numbers.

Rational numbers were first defined as *those numbers which can be expressed as the ratio of two signed integers*, and were the next logical extension of the number set after introduction of negative numbers.

They provide all the fractions which can be written, and include all the signed integers because the denominator (or lower number) of the ratio can be the signed integer +1, giving the rational number 2/1 all the properties of the integer 2.

The simple definition of rational numbers implies that 4/2 is a different number from 2/1 or 8/4. These ratios all have identical qualities, however, so the definition was extended to require that the ratio be expressed "in lowest terms." That is, the integers which form the ratio may have no common factor. Those numbers formed by a ratio of integers which do have common factors are simplified by removing the common factors, which reduces them to lowest terms.

With this extension of the definition, the ratio 4/2 is seen to actually be $(2 \times 2)/(1 \times 2)$, and removing the common factor of 2 from both parts of the ratio leaves the simpler 2/1.

Frequently we will find it convenient to write rational numbers in other than their lowest terms. If we want to express the ratio 3/2, but write it as a decimal fraction, we first introduce the common factor 5 to both parts to change the expression to 15/10, then write it with a decimal point as 1.5. For all definitions, though, the lowest-terms representation of a rational number is used.

It might seem that with the introduction of fractions, all possible numbers had been defined—in fact, only a tiny part of the possible numbers are included in the set of rational numbers. The set of rational numbers does, however, include every number which can be represented with total precision. Any representation of a number which is not a member of the rational set can never be completely accurate.

The problem with rational numbers was that, although they could approximate any quantity as closely as desired, they are inevitably separated by gaps; rationals are defined as ratios of integers, and integers are whole numbers which have gaps between them.

By making the integers we use large enough, we can make the gaps in the series as small as we like—but we can never eliminate them. Those numbers which fall in the gaps are sometimes called *irrational* numbers because they are not rational.

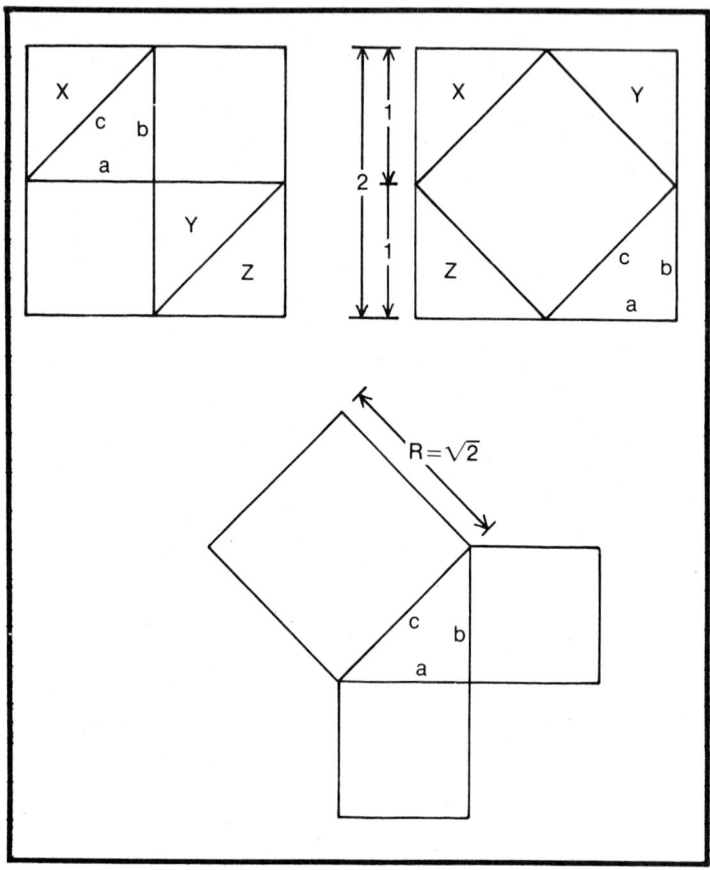

Fig. 3-6. Pythagoras' proof that square erected on hypotenuse (c) of right triangle is equal to squares on its sides (a, b) is shown here. Small squares in upper left figure are one unit on each side, but have same area as single square in upper right figure. It's length is not a rational number.

One of the first irrational numbers discovered was the square root of 2. This is the number which, when multiplied by itself, gives exactly 2 as a product.

Pythagoras proved that the square of the hypotenuse of a right triangle was equal to the sum of the squares of its sides. When a right triangle having both its sides equal to 1 is constructed, the square of its hypotenuse must then be 1+1, or 2 (Fig. 3-6).

But no rational number exists with which to measure the length of this hypotenuse. We know that the length is more than 1, and less than 2, since 1 × 1 = 1 and 2 × 2 = 4. We can find by trial and error that it is larger than 1.4 and smaller than 1.5. since $1.4 \times 1.4 = 1.96$ and $1.5 \times 1.5 = 2.25$.

As we continue our trial-and-error search, we find that the ratio 707/500 (1414/1000) is too small and 283/200 (1415/1000) is too large. No matter how far we go, we only approach the number we are seeking. Either we never quite get to it, or we overshoot and go too far.

If we abandon our trial-and-error method, and turn to logic as the Greeks did, we can prove that no such ratio can exist. Let's call the unknown number we're seeking R, so that $R \times R = 2$.

If R is a rational number, it must (by the definition of a rational number) be expressible as the ratio $R = I/J$, where I and J represent two integers which have no common factor.

When we substitute I/J for R in the statement $R \times R = 2$, and do the arithmetic, we find that we have a new statement: $(I \times I)/(J \times J) = 2$. This can then be rearranged into the statement $(I \times I) = 2 \times (J \times J)$.

This last statement implies that I must be an even number since it is stated to be equal to two times the quantity $J \times J$, and an even number is one which has two as a factor. If I is even, it is divisible by 2, so $I \times I$ must be divisible by 4 (which is 2×2).

But if $I \times I$ is divisible by 4, then $2 \times (J \times J)$ is also divisible by 4; setting up this division and factoring gives us the statement that $J \times J$ is divisible by 2.

We have just proved that both I and J have the common factor 2, if they exist, but the definition of rational number requires that the two integers involved not have any common factor. Assuming the existence of a rational number representing the square root of 2 leads us to a contradiction; we know that no such number can exist.

Since the number itself does exist—the hypotenuse obviously has a length—and yet cannot be expressed as a rational number, we know that gaps must still remain in our sets of numbers.

Continuous Numbers

To change our numbering schemes from ones that mark only discrete points (such as the integers and the rationals) into those which can describe and identify continuous lines, we must introduce the idea of comparison.

When we measure out a pound of butter, we do so by comparing the weight of our butter to some predetermined standard. If we have too little butter, we add more; if too much, we take some away.

Similarly, to identify a number which is not in the set of rational numbers, we compare it to the rational numbers. As we carry out this comparison for an unlimited period of time, getting closer and closer to the goal with every attempt but never quite reaching it, we are slicing the set of rational numbers into two subsets. One of these contains only numbers which are smaller than the irrational number involved. The other contains only numbers which are larger.

Since the set of integers with which we construct the ratios to be compared is itself unlimited, the comparisons can never reach a limit. Each comparison, though, will narrow the gap within which the irrational number lies.

We can imagine such a process going on forever, and use this "cut" of the rationals into two subsets to mathematically define the irrational number as being *that number which remains in the gap when the set of rational numbers has been completely divided.*

To put it a little more plausibly, we can determine that *if* we can find any way to separate the set of rational numbers into two subsets, called L for *less* and G for *greater*, then one of two situations must be true: Either there is a rational number which separates the L subset from the G subset, or a new number can be defined to mark the point of partition.

For each irrational, we then define a *real number* to be such a partition, which leaves every member of the L subset smaller than every member of th G subset.

A system of this sort includes all the irrational numbers (such as the square root of 2), as well as the *transcendental* numbers (such as pi, the ratio of the circumference of a circle to its diameter), epsilon (the base of natural logarithms), and phi (the Golden mean).

The system of real numbers defined by partition of the set of rational numbers includes a separate number for every point on a line, and leaves no room for extension while maintaining the idea of order or sequence. However, one more extension turned out to be necessary in the development of mathematics.

When you take the square root of a *positive* number, it turns out to be a member of the set of real numbers; however, when you take the square root of a *negative* number a problem arises.

The problem is based in the rule for multiplication of signed integers, which requires that the product of two negative numbers be positive. The only way a product can be negative and still obey this rule is for its factors to have opposite signs. Yet a square root is, by definition, a factor which when multiplied by itself yields the square as product.

For instance, $+2 \times +2 = 4$, and $-2 \times -2 = 4$. Thus, both $+2$ and -2 are separate square roots of 4. The only way to get -4 by multiplying twos is $+2 \times -2 = -4$, and this involves two different factors (differing by the sign involved).

The stroke of genius which solved the problem was the recognition that -4 could first be factored into $-1 \times +4$, so that the square root of -4 became the same as the square root of -1 times the square root of $+4$.

The mysterious square root of -1 was just as elusive as ever, but it had become a single anomaly in the system and so could be singled out or special treatment. It was named the *imaginary number* and given the symbol i.

This made it possible to write the square root of -4 as $2i$. While it didn't shed any light on how to assign a place in the sets of numbers to i, it did make it possible to use i as an operand.

Subsequently, physicists discovered a real meaning for the idea represented by i, and it's now a key part of the mapping of mathematics to the real world—but that's getting ahead of ourselves.

Number Lines and Planes

When we extended the set of rational numbers to obtain the set of real numbers, we found that we had enough numbers to provide a different one for every different point on a line.

The image this brings to mind is one widely used in the study of mathematics, and it even has a name—the *number line*.

To draw a number line, you simply draw a straight line and mark a point near its middle. Call this point *zero* (Fig. 3-7).

Now measure off equal distances starting from this point, moving to the right. If you make the distance one inch, you'll mark a point one inch from zero, then one two inches away,

Fig. 3-7. **Number line** is simply a straight line with zero point indicated on it.

and so forth (Fig. 3-8). The points you just marked correspond to the members of the set of natural or cardinal integers, and since you can imagine that your line extends as far as you want it to, the set has no limit.

These same points correspond to those members of the set of signed integers with have positive signs. Marking similar points, but moving left from zero instead of right, gives you points which correspond to the negative integers (Fig. 3-9). The line's length is unlimited in this direction also.

The points we have so far marked provide a mapping for the complete set of signed integers, but also for those members of the set of rational numbers which have +1 for their denominator, such as $-2/+1$, $-1/+1$, $+1/+1$, $+2/+1$, and so on.

To provide points to correspond to the rest of the rational numbers, measure off distances which have the corresponding ratios. If the distance between the points representing 0 and 1 is one inch, then the rational number 1/2 will be represented by the point at which 1 inch is split into 2 parts, 1/2 inch from zero. Similarly, the number 3/2 is represented by the point 1½ inches from zero (where 3 inches is split into 2 parts), 3/4 by the point 3/4 inch from zero, and so forth (Fig. 3-10).

There's another way to look at the same thing: Consider the denominator (the lower half of any fraction) to be the number of parts that make up one inch (in this case, since we've split our number line into inches). So 3/4 not only means "3 inches split into 4 parts," but it means "three of the four

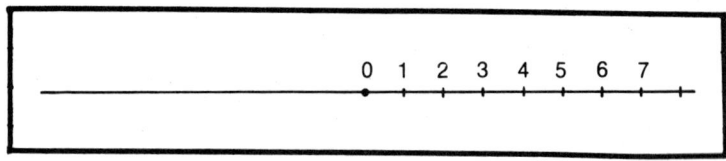

Fig. 3-8. Counting numbers or cardinal integers extend to right at regular intervals along number line. Each interval represents an increment of one unit.

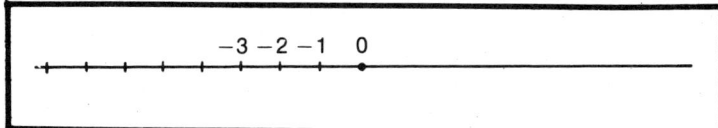

Fig. 3-9. Negative numbers extend to left along number line. Only difference between positive and negative numbers, in this analogy, is the direction of increment.

parts we've split an inch into." And 17/25 would mean 17 of the parts or increments we've split an inch into.

Since the number of points on any line segment is unlimited, we could map the entire set of rational numbers into the segment of the number line between *zero* and *one* and still have points left over. These leftover points represent members of the set of real numbers.

Besides helping us visualize the differences between cardinal integers, signed integers, rational numbers, and real numbers, the number line assists in the study of addition and subtraction.

Addition of positive integers, or counting up, is equivalent to moving in the positive direction, or to the right, on the number line. If we have marked the integers at one-inch intervals on the line, adding 3 to 5 means that we start at the point marked 5 and move 3 more integers to the right, which brings us to the point marked 8.

Note that this is not a definition of addition, but an illustration of it. We'll get into more detail in Chapter 5.

Subtraction, in contrast to addition, is equivalent to moving in the negative direction, or to the left, on the number line. Subtracting 3 from 5 means that we start at the point marked 5, as before, but move 3 integers to the left instead of to the right, ending on the point marked 2.

The number line also shows us that things turn around when we pass through zero in the negative direction to reach the negative numbers.

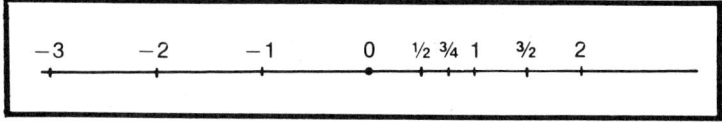

Fig. 3-10. Rational numbers fit onto number line in spaces between integers as shown here: ½ is halfway between 0 and 1, ¾ is halfway between ½ and 1, and so forth.

For instance, with positive numbers the digits are in the sequence 1, 2, 3, 4, and so on as we move to the right.

With negative numbers, moving to the right moves us past the digits in the sequence $-4, -3, -2, -1$.

Since adding positive integers is equivalent to moving to the right, adding $+2$ to -4 moves us two integers to the right from a starting point of -4, bringing us to -2.

Subtracting $+2$ from -4 would move us two integers to the left, bringing us to -6.

Going back to our addition example, it appears that so far as the digits are concerned, adding two numbers with different signs amounts to subtracting the smaller from the larger and taking the sign of the large.

Adding -2 to $+4$ would, by this rule, give us $+2$. This, however, is equivalent to moving to the left along the number line. Addition of negative numbers, we see, is equivalent to moving to the left.

On the number line, *positive* and *negative* are directions; adding a number means to move in its direction, and subtracting it means moving in the other direction.

We'll get into more applications of the number line in later chapters. Right not, let's see how the idea of the number line makes it possible to visualize numbers which include that imaginary factor, *the square root of -1*.

We have established that the number line provides a one-to-one correspondence between the points on the line and the set of real numbers.

We also discovered that all the imaginary numbers could be considered as products obtained when we multiply the real number expressed with the same numerals, by the operand i which represents the square root of -1.

Thus, we can establish a one-to-one correspondence between the points on a second number line and these numbers, with the difference between the two number lines being the fact that one represents numbers multiplied by i, and the other represents real numbers.

Either of our number lines can be drawn on a flat surface if it is large enough. Such a surface we call a *plane*. In addition, both lines have one point in common. That's the point named *zero*, because in any kind of numbering, zero always means *the number of elements in the empty set*, or *nothing*.

This common point at zero means that if we draw both lines on the same plane, they will cross at the point named *zero* for each line. Since the lines must cross, they will cross at some angle.

Quick subject change: Stand in a relatively open space, facing some object (or imagine that you're doing so). Now do what the military folk call a *left face* movement. Do it again, a third time, then once more, so that you have executed the left-face motion four times, and youll find that you have turned completely around so that you're once again facing the same object you were at the start.

Try it again, standing on a line and facing one end of the line. This time do the left face only twice, and you'll find that you're looking at the other end of the line. If you were looking at the *right* end at the start, you're now looking at the *left*.

When we were examining the use of the number line to illustrate addition and subtraction, we found that the difference between positive and negative mapped into the difference between left and right on the number line.

Doing the left-face maneuver twice corresponds to a complete turn, from facing right, to facing left. Doing it four times corresponds to a full rotation, from facing right, through facing left, back to facing right again.

Since i is defined as the square root of -1, i squared or $i \times i$ must equal -1. Since -1×-1 equals $+1$, we can do a little substitution to show that $i \times i \times i \times i$ also equals $+1$.

If we let multiplication by i be equivalent to doing a left face at zero on the number lines, what happens?

Let's imagine that we are standing at the zero point of a number line, facing along the line toward positive infinity. If we now do a left face, we could be looking down a second number line (the one multiplied by i), toward its positive infinity. (See illustration, Fig. 3-11.)

A second left face, corresponding to a second multiplication by i, would bring us back to the real number line, but we now would be facing toward negative infinity. This confirms that two left faces on a number plane correspond to $i \times i$, or -1.

The third left face would then bring us to the negative side of the imaginary number line, confirming that $i \times i \times i$ is equivalent to $-i$, and the fourth would return us to the positive real line, confirming that $i \times i \times i \times i$ corresponds to 1.

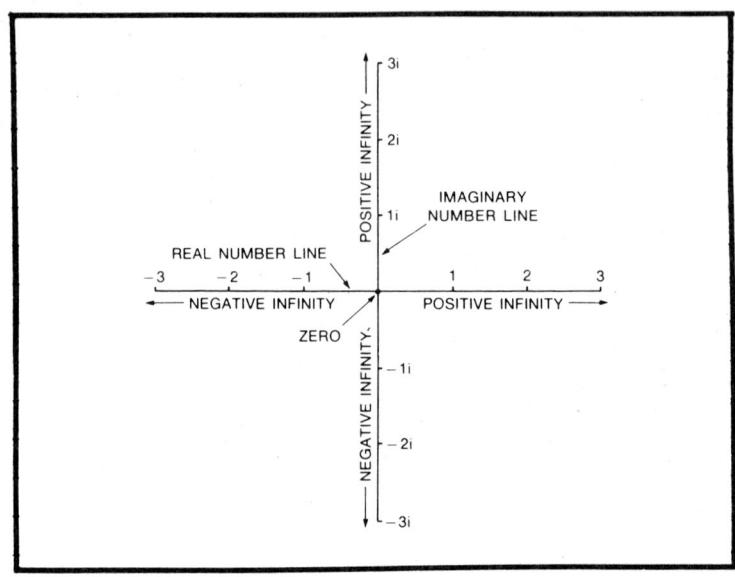

Fig. 3-11. Placing two number lines at right angles to each other, intersecting at their zero points, gives us the number plane which provides a point on its surface for every one of the unlimited quantity of complex numbers. Quantity **i** represents **square root of -1**, the "imaginary" number which has turned out to be so useful in electronics and physics.

This shows that if our real and imaginary number lines cross at right angles to each other on the number plane (Fig. 3-11), then a 90-degree rotation (or right angle) on that plane will be equivalent to multiplication by i.

It does more than that; in addition to the unlimited number of real numbers corresponding to points on the real number line, and the unlimited number of imaginary numbers similarly represented by the imaginary number line, we now have all those points on the plane which are not on either line, which can also represent numbers.

To account for these possible numbers (which turn out to be of utmost importance in most applications of math to the real world of science), we define the set of *complex numbers*. Every complex number can be described by two parts, one real and one imaginary. Where a real number might be represented as 3.14159, and an imaginary as $2.128i$, a complex number is represented as the sum of its parts: $3.14159 + 2.128i$.

Just as every cardinal number has a corresponding member of the set of signed integers, every signed integer has a corresponding member of the set of rational numbers; and just as every rational number has a corresponding member of the set of real numbers, so does every real and imaginary number have a corresponding member of the set of complex numbers.

This means that the set of complex numbers includes all known numbers. The real number 3.14159 would be represented in complex form as $3.14159 + 0i$, and the imaginary $2.128i$ would be represented as $0 + 2.128i$.

The number plane and the set of complex numbers give us a point for every different number used in the real world. Mathematicians may not, however, be through expanding the sets of numbers; all we can say for sure is that they haven't done more yet.

Accuracy and Precision.

It's unfortunate that the names *real* and *imaginary* were applied to the sets of numbers which went into the development of the set of complex numbers, because neither of them is any more or less "real" than the other.

Both, in case it slipped past you in the explanation, are purely mathematical concepts which have no exact counterparts in the real world.

As we extended the concepts of numbers from the simple counting numbers, the cardinal integers, into the complex number plane unlimited in any direction, we moved ever more distant from reality toward the abstract, completely fictional universe of pure mathematics.

The final departure from reality occurred when we moved past the set of rational number. Any members of the set of rational numbers can be written with complete precision. If we write the number 159/200 it means exactly that. Since (by definition) all rational numbers with common factors above and below the bar represent the same number, our 159/200 is the same as 159000/200000, but it is totally different from 158999/200000, or 159001/200000.

That is, the simple act of writing a rational number defines that number completely; there is no room for error.

A real number, on the other hand, cannot be written at all unless it happens to correspond exactly to some rational number—and even then we cannot show merely by the way it is written that it does correspond to the rational we use to write it.

If we write 159/200 to represent a real number, all we are saying is that our real number is something larger than 158/200 or 159/201, and smaller than 160/200 or 159/199. We may be representing the number 159001/200000 or 158999/200000, and no one will ever know.

This is more apparent when we write the rational approximation to the real number in decimal form. The fraction 159/200 we have been using becomes 0.795 when we convert it to decimal form (795/1000).

But if it is approximating a real number, the actual number could be anything from 0.794500000000000001 up to 0.795499999999999990 (we stopped at 18 decimal places only because we had to stop someplace—actually, an unlimited number of additional digits could be put in as well), if the approximation is rounded off to "three places" by conventional rules.

The whole idea of rounding off applies mainly to real numbers and their corresponding abstractions, the imaginaries and complex numbers. It's based on the fact that no real number can be accurately represented in a limited number of digits, so it doesn't hurt to introduce a little more error by cutting things off short.

If the so-called real numbers and their brethren are only figments of the imagination, why bother with them? The major reason is that the real world is equally limitless. While we can count eggs, we have to measure butter—and the only way we can measure anything is to compare it to some standard.

The comparison process which we use to measure anything is exactly like the comparison of real numbers to rational numbers, which we discussed in extending the set of rationals to become the set of reals. We can never know for sure that we have exactly matched the standard, but by taking enough care we can make the error just as small as we please.

This brings us around to the difference between accuracy and precision, which many persons often find confusing. The term *precision* refers to the number of digits we choose to use to represent a real number, or the parts of a complex number. *Accuracy*, on the other hand, indicates how small we have made the error when we measure something, or approximate a real number by a rational.

If we measure something to an *accuracy* of plus or minus 1%, this means we have kept the error below one part per hundred. If our measurement comes out as 72 inches, we know the error is not greater than 0.72 inch; we do *not* know how great the error is, or whether it is in the direction of being above or below the actual length (if we know it, it wouldn't be error).

In the preceding paragraph, we expressed the figures to a *precision* of two digits or places. To increase the precision to three places, we would simply say that we measured a length of 72.0 inches, to within 1% accuracy. The actual error would not change; it would still be somewhere between -0.720 and $+0.720$ inch. The difference in precision made no change at all in the accuracy.

That's the point which brings in the confusion. If you measure a piece of lumber as being 12½ inches wide by 35⅝ inches long, using a pocket rule with sixteenth-of-an-inch gradations, the accuracy of each measurement is within a sixteenth (0.0625) of an inch.

If we then calculate the area using the greatest precision we can achieve (and since the fractions are rational, we can achieve total precision), we will multiply 12.5 times 35.625 to find that the area is 445.3125 square inches.

The *precision* of our calculated result is seven digits. Either of our two measurements, however, could have been 1/16 inch off in either direction, so the *accuracy* is less than the calculation would indicate.

If both measurements erred on the side of being too big, the actual area could be as small as 442.30859375 square inches. If both measurements were too small, the figure would be 448.32421875. The *usable* precision of our result, because of this lack of accuracy, is only two digits: 445 square inches, give or take 3 inches either way.

In general, the usable precision of the result of any operation is no greater than that of the least precise or least accurate operand involved. In this case, it was the smaller of the two measurements, since the error (rather than the percentage of error) remained the same.

The rise in availability of the pocket calculators, which always provides the result to the greatest precision of which it is capable, may perpetuate the confusion between precision and accuracy. The thing to remember is this: while precision is required, any more of it than you have to use to show the figures accurately tends to confuse rather than to help.

Chapter 4

Relations, More or Less

We've seen that mathematics is a language which makes statements about an ideal universe which is limited by agreement at the start of any discussion, and in which all the rules are always obeyed.

These mathematical statements are always cast in the form of statements expressing relations between operands; the rules of the universe then determine whether each statement of relation is true or false.

Before we can make much use of the language, we have to know what all these words mean. Earlier we found meanings for operands and learned that in arithmetic, at least, they usually turn out to be numbers. The previous chapter told us more than we possibly really wanted to know about numbers.

The next step, then, is to discover what we mean by *relations* and by *true or false* in the language of mathematics. When we've accomplished that, we'll see how relations fit into the general picture of arithmetic and everyday usage.

RELATIONS—WHAT ARE THEY?

When we examined the concept of number, we found that it's an abstraction of the idea of quantity. That is, it's determined by the content of any given set being discussed.

Our first definition for number, in fact, was as the property held in common by several sets, which had as their only common property the fact that each contained the same quantity of elements.

The idea of relation is fundamentally different. Where number is an idea related to content of a set, *relation* has almost nothing to do with content. Instead, it's a statement about the patterns formed by the elements.

In this section, we'll attempt to define the concept of relation in its mathematical sense, then describe the properties by which relations can be described, and finally define the relations most commonly used in everyday math.

Fencing In the Relations

You'll recall from Chapter 2 that a definition is simply a fence around an idea, to separate that which is essential to the idea from that which is not.

The idea to separate, which we used in that statement, could be considered as a relation, since separation is essential to the formation of a pattern from an set of elements.

Not all relations, though, are simple separations. A relation may appear in many forms. The ones we deal with most often in math are called *dyadic* or binary relations. This doesn't mean that they apply only to the binary number system we met in Chapter 3; rather, it indicates that they describe the connection (or lack of same) between exactly two operands.

We found that we could generate any cardinal integer by starting at unity and applying the successor function again and again.

Similarly, we can handle relations of any reasonable number of operands by taking them two at a time, since a mathematical statement may deal with any number of relations at once.

That successor function, incidentally, gives rise to a relation which is a little more complicated than simply one of separation. We can call it *being the successor of* and describe it by the assertion that one of its operands is the result of applying the successor function to the other.

By changing one word of the name, we can change the concept rather dramatically and describe a different relation: *being a successor of*. This asserts that the first operand is either the direct successor (as in our previous example), or a remote successor (two or more applications of the function

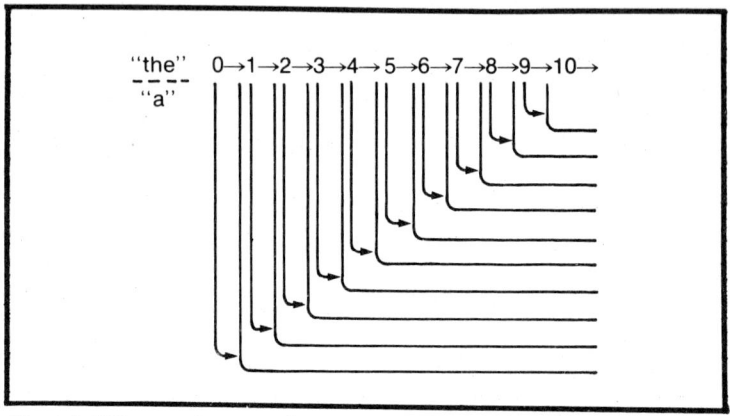

Fig. 4-1. Difference between **the** successor and **a** successor of a number is shown here. All of the numbers shown, except 0, are in the set **a successor of zero**; only 1 is **the** successor of zero. Sequence extends to the right without limit.

away from) of the second operand. Figure 4-1 shows this distinction visually.

Thus, "5 is the successor of 4" is a true statement by general standards, as is "10 is *a* successor of 2." Neither "4 is *the* successor of 5" nor "10 is *the* successor of 2," however, would be considered true.

Most relations have *inverse* relations associated, which reverse the direction of the relationship. The inverse to the *a successor of* relation is often called the *ancestor* relation: "4 is an ancestor of 5" would be illustrative of this usage. If the idea of parentage in numbers disturbs you, call it *predecessor* instead.

Any statement describing the connection between two elements of the same or different sets can be considered as a statement of relation.

Consider, for instance, these examples from fields not usually thought of as mathematical: a brother of...., a parent of...., a child of...., a resident of...., a student of.... This list could go on without end; the real world abounds in relations.

Any definition which uses the idea being defined as an essential part of the definition itself is usually considered faulty and of little use. Philosophers call such a tactic a *circular* definition. Nevertheless, the temptation is strong to declare that a mathematical statement of relation is a statement which describes a relationship existing between the elements to which it applies. (The word *relationship* is used here in its everyday real-world sense.)

That is, the relation may be a statement of sequence, if sequence is the property of the relationship which is being described. It may be a statement of identity, if that is the property under discussion. Or it may be a statement of any other imaginable relationship which can exist between its elements.

Every definition, then, can be considered as a statement of an equivalency relation, in that it defines a name or label as being equivalent to or interchangeable with a set of ideas.

In giving examples drawn from the previous chapter, the close association between the idea of *function* (which has, so far, appeared only as the arbitrary and undefined *successor function*) and that of *relation* (which we are here attempting to adequately fence in) was exposed.

In Chapter 2, we learned about *mapping* of one set into another. A *function* is like mapping, and so is a *relation*; both involve selection of subsets from a product set.

In both cases, the product set is that formed by taking the product of a set and itself.

A relation is similar to a mapping when the elements to which the relation may be applied are defined as a set; this sentence can be considered as such a definition, since every element to which the relation applies is an element of the set under discussion.

The two elements which are made part of any statement of relation, then, must form a specific subset of the product set. Since relation is a statement of pattern, the subset is ordered, and forms an ordered pair. The relation itself is then defined much as we did *number*; it is the property common to all such ordered pairs.

You can see that the relation, in itself, becomes a description of the subsets. Similarly, a mapping was found to be a specific subset (also from a product set) which matched the real world.

Let's go back for a moment to our mapping example of houses and children. We'll give labels to a couple of relations, without formally defining them, so we can use them as examples. When we write *a lives in b*, we are saying that in the real world, child a is a resident of house b; the inverse, *b is the home of a*, says that house b is the residence of child a.

If we have three houses—brick, frame, and stucco—and six children—Jack, Tom, Jill, Jane, and Nancy—we might have mapping such as: Jack/brick, Tom/brick, Bob/brick, Jill/brick, Jane/brick, Nancy/brick.

This is the same thing as the following group of statements of relation: Jack lives in brick. Tom lives in brick. Bob lives in brick. Jill lives in brick. Jane lives in brick. Nancy lives in brick.

Another mapping from the examples in Chapter 2 was: Jack/brick, Tom/brick, Bob/frame, Jill/stucco, Jane/frame, Nancy/stucco.

The corresponding statements of relation are: Jack lives in brick. Tom lives in brick. Bob lives in frame. Jill lives in stucco. Jane lives in frame. Nancy lives in stucco.

The statement of inverse relation for this situation becomes: Brick is the home of Jack. Brick is the home of Tom. Frame is the home of Bob. Stucco is the home of Jill. Frame is the home of Jane. Stucco is the home of Nancy.

Thus, every statement of relation we have made here is equivalent to either a partial mapping of a set into itself, or the inverse of such a mapping. As we proceed, we shall see that this is true for all relations, and so can be used as a definition or fence for the idea of *relation* in math.

Properties of Relations

Now that we have established a relation as being a special kind of mapping, let's see what properties all binary relations have in common so that we can classify them. To pinpoint these properties, we'll work with relations drawn from family life, rather than working with numbers.

For a start, let's try *is the parent of* as a relation. If the statement *John is the parent of Debbie* is true, then the statement *Debbie is the parent of John* (Fig. 4-2) cannot be if it's the same Debbie and John in both statements.

Apparently the relation *is the parent of* must have some different relation as its inverse; we call this inverse *is the child of*; and in our previous example, *Debbie is the child of John* would be true.

Not all relations require separate inverses. Some, in fact, are their own inverses. For instance, *Debbie is the sister of Joan* is the inverse of *Joan is the sister of Debbie*, and if one of

Fig. 4-2. If **John is the parent of Debbie** is true (top), then **Debbie is the parent of John** cannot be true at the same time (bottom). This illustrates a relation which is nonsymmetric.

these statements is true then so is the other. The relation is the same in both, however.

The relation *is the sister of* is not always its own inverse, because it's not true that *Robert is the sister of Debbie* even if *Debbie is the sister of Robert*. The relation is freed from its sex label by using the psychologists' word *sibling* to replace both words *brother* and *sister*, to produce a relation which is consistently its own inverse.

In mathematics, we find some relations which are their own inverses, and some which are not. This is one of the three major properties by which we classify relations. A relation which is its own inverse we call *symmetric*, and one which is not is called *nonsymmetric* or *asymmetric*.

Thus, the relation *is the parent of* or *is the child of* is asymmetric, while *is the sibling of* is symmetric.

To define symmetry of a relation formally, we use the symbols a and b to stand for the elements, and * to stand for any relation. We then assert that if $a*b$ implies $b*a$ (that is, if $b*a$ is true whenever $a*b$ is), the relation * is symmetric; if not, is asymmetric.

In addition to the relation *is the parent of*, we can define the relation *is an ancestor of*, just as we did for the successor function in numbers. Thus if *John is the parent of Debbie* and *Debbie is the parent of Carl* are both true, then *John is an ancestor of Carl* would also be true.

Note that *John is an ancestor of Debbie* would be true as well, since the parent is simply the closest ancestor in the lineage.

This leads to the observation that if *John is an ancestor of Debbie* and *Debbie is an ancestor of Carl* are both true, then *John is an ancestor of Carl* is true also.

Here we have found another property of relations, totally different from the property of symmetry. This property is called *transitivity*, and it can be called a property of repeated applications of the same relation. That is, if the relation is true for the ordered pair of John and Debbie and also for the ordered pair of Debbie and Carl, then it is true for the ordered pair which results if Debbie is eliminated from both—the pair (John, Carl). It's an elimination of the middle.

Formally, using the same symbols as before, if $a*b$ and $b*c$ being true together imply that $a*c$ is true, the relation * is said to be *transitive*; otherwise, the relation is nontransitive. This is the second of the three major properties of relations.

So far, the family relationships we have used for examples have been parent—child or brother—sister relationships. The third property of relations is more difficult to illustrate clearly with personal relationships. One family relation which exhibits the property, though, is *has the same name as*.

Within most families, the only person having the same name as any member is that member in person. That is, *John has the same name as John* is true, but *John has the same name as Debbie* is not.

This property, defined formally with our usual symbols as the case in which $a*a$ is true for every element of the set, is called being *reflexive*, and relations which do not have it are called nonreflexive.

These three properties—reflexivity, symmetry, and transitivity—are used to define all types of mathematical relations.

For example, any relation which possesses all three of the properties is called an *equivalence* relation, and implies that its elements may be freely interchanged without affecting the characteristics described by the relation (the property of interchangeability is usually implied by reflexivity alone).

Undoubtedly, the most familiar equivalence relation is that called *equality* and denoted by the *equals* symbol, $=$. All of arithmetic, and most of algebra, is based upon this relation.

It is not, however, the only equivalence relation. Let's look at another candidate—the relation *has the same last name as*.

John Jones has the same last name as John Jones is true, so the relation is reflexive. *John Jones has the same last name as Debbie Jones, so Debbie Jones has the same last name as John Jones* is also true, proving the relation to be symmetric. And finally, *John Jones has the same last name as Debbie Jones, and Debbie Jones has the same last name as Carl Jones, so John Jones has the same last name as Carl Jones* is true, to make the relation transitive.

And since the relation is reflexive, symmetric, and transitive, it must (by definition) be an equivalence relation. So far as having the same last name is concerned, one *Jones* is interchangeable with any other.

Note well that this relation says nothing about equivalence of any other characteristics of those to which it is applied; its universe of discourse is limited to the single property of last names.

An equivalence relation is simply another name for the division of a set into subsets. You can see from the previous example that the equivalence relation *has the same last name as* can be used to separate out the subset *people named Jones* from the set *people*, and it's only a short step from this to the idea that the same relation can be used to select any subset of people by last name.

This equivalence relation corresponds to a partition of the set of *named people* into subsets, each subset consisting of those having the same last name.

We'll run into this idea of *set partition* and *equivalence relations* being the same idea expressed in two different ways again. Right now, let's move along to meet the more frequently used arithmetic relations.

Commonly Used Relations

In arithmetic, we commonly use half a dozen relations, although we seldom label them as such. The most common of these is the relation *equality*, which we denote by the symbol $=$.

Equality is an equivalence relation, as expressed by the rule that *things equal to the same thing are equal to each other* (a rule, by the way, which is true in arithmetic but not necessarily in all the rest of mathematics).

As an equivalence relation, equality is reflexive, symmetric, and transitive. That is, for any number n, all the following statements are true: $n = n$, if $m = n$ then $n = m$, and *if* $x = y$ *and* $y = n$ *then* $x = n$.

Most of us met the relation of equality at the same time we met addition; the statement was written $1+1 = 2$ and read as *one and one are two*. Once we accepted that idea, we went on to use it throughout all the rest of the math we studied and made use of, probably without questioning its logic.

In addition to equality, two other relations (Fig. 4-3) are often encountered. Both of them deal with the idea of sequence; one is called *greater than* and symbolized by an arrowhead pointing to the right ($>$). The other is called *less than* and is symbolized by a similar arrowhead pointing to the left ($<$).

While these relations are similar to each other, they are neither identical nor exact opposites, as we shall see shortly.

Both are transitive, but neither is reflexive or symmetric. That is, if we use the symbol .gt. for *greater than* and .lt. for *less than*, for any number *n* the statements *n*.gt.*n* and *n*.lt.*n* are both false (nonreflexive), and similarly false are the statements *if n.gt.m then m.gt.n* and *if n.lt.m then m.lt.n* asymmetric).

Both the relations specify relative locations of their operands in an ordered set, much like the parent and child relations in our earlier family examples.

The *greater than* relation asserts that its first operand appears later in the set's sequence than does its second operand. The *less than* relation reverses the sequence to assert that the first operand appears earlier than the second.

While this might give the impression that the two relations are complementary one to the other, or are inverses, such is not actually the case. If one operand is greater than another, the complementary situation would be for it to be *not* greater, which is the same as being *equal to or less than*.

The set of six common relations is, in fact, composed of the three we have already met, plus their complements. For instance, the relation *unequal* or *not equal* is the complement of equality, and asserts that the equality relation does not exist between its operands. It says nothing else about relative sequence.

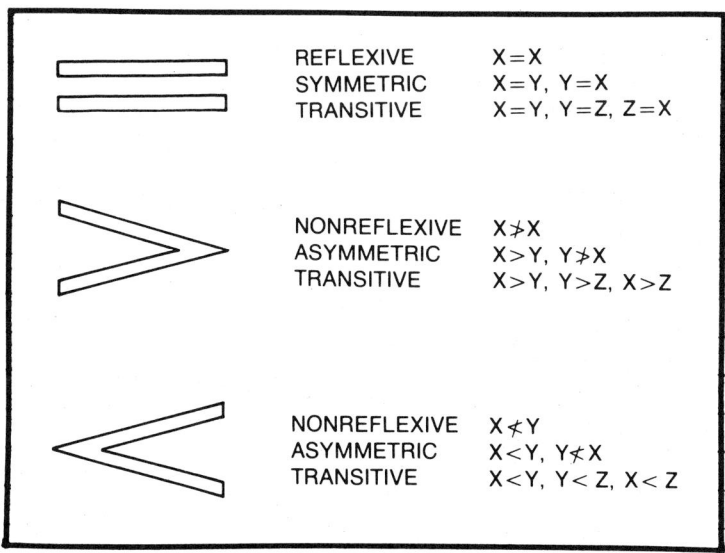

Fig. 4-3. Distinguishing characteristics of three basic relations of arithmetic are listed here. Only "=" is reflective, symmetric, and transitive. The other two are only transitive.

However, if either the *greater than* or *less than* relation holds between any pair of operands, the *unequal* relation is also true of that same pair.

Similarly, if the *equal* relation is true of any pair, neither *greater than* nor *less than* can be true of the pair in any sequence, since both .gt. and .lt. are nonreflexive relations. Nothing can be greater or less than itself.

The complement of the *greater than* relation is, as we noted previously, *not greater than* or *less than or equal to*. If *a* is not greater than *b*, this does not exclude the possibility that they are equal. If they are not equal, then *b* must be the greater.

By similar reasoning we find that the complement of the *less than* relation is *not less than* or *greater than or equal to*.

The *not greater than* relation can be used to partially order any set of numbers by applying the relation repeatedly to determine the smallest number of the set, then partitioning the rest of the set out as a subset and repeating the action on the subset, until every member of the set has been assigned a position in the sequence.

It's called a *partial ordering*, since nothing in this procedure excludes the possibility of having several *equal* members in any of the subsets. If this happens, no order is established among the *equal* members even though they are, as a subset, assigned a position in the sequence.

Formal definition of the *greater* and *less* relations is based upon the successor function that establishes the set of natural numbers. Repeated application of the successor function to each new number generated leads to the idea of the *ancestor* of any number; each number is the *ancestor* of its successor, and of all successors of its successors.

With that concept established, the *less* relation is defined as follows: M is less than N if and only if M is an ancestor of N.

Once we have defined the *less* relation, we can use it to define *greater*: X is greater than Y if and only if Y is less than X.

Since the equality relation is formally defined by the properties of relations, all three basic relations are now defined formally.

For any two numbers in the set of integers, one and only one of these three relations can be true. That is, any integer picked at random is either less than, equal to, or greater than a second integer also picked at random. This rule is known as the *trichotomy law* since it cuts the set of numbers into three parts.

The other three relations are formally defined as the complements of the three already defined, and for them the complement of the trichotomy law is true: for any two integers, one and only one of these relations is false, leaving the other two true.

Thus, 1 is less than 2, not equal to 2, and not greater than 2. Similarly, 3 is not less than 3, it is equal to 3, and it is not greater than 3. Finally, 5 is not less than 4, not equal to 4, and it is greater than 4.

If you conclude from this that only three of the six relations are necessary, you're absolutely right. All six, however, are in common use despite the fact that the *complement* set duplicates the meanings of the *basic* set. In the real world, we use whichever of the six best fits the problem at hand.

WHAT IS TRUTH?

Nearly two thousand years ago, Pontius Pilate asked the question "What is truth?"

To this day, no satisfactory answer has been found in the real world of everyday life. In mathematics, however, which as we have seen has only a tenuous connection with the real world, the idea of truth is a simple concept and easily defined. This fact, alone, makes possible most of the rest of mathematics.

Mathematical Truth

In mathematics, *truth* is defined as *the absence of contradiction*. This may, at first, sound like doubletalk; if it strikes you as such, try an alternate pair of statements:
1. *Any statement about relations that contradicts itself is false.*
2. *Every statement that is not false is true.*

Our first definition simply summarizes the alternates, and asserts that every statement about relations which does not contradict itself is true in the mathematical sense.

For the phrase *true in the mathematical sense* to have any meaning, the statement must of course be a mathematical one. *You are a crook* is a relational statement, about an implied equivalence relation, and does not necessarily contradict itself. Thus, it is a candidate for being called *true*.

It is not, however, a mathematical statement. This fact disqualifies its candidacy, and means that our mathematical definition of *truth* does not apply.

The similar statement *Either you are a crook or you are not* comes closer to being mathematical, but until the properties which establish membership in the set of all crooks are strictly defined it says nothing about any individual, nor is it truly mathematical. Its truth, therefore, remains undefined.

Defining things in terms of what they are not, while a valid technique (remember that any set, once defined, also defines its complement as all the rest of the universe), often turns out to be a confusing way of doing things.

Let's take a positive viewpoint instead, and see if we can show by example what constututes mathematical truth.

Mathematics, remember, makes statements about relations between operands. We have examined the properties of the relation called *equality*, and found that it means that the two operands are interchangeable one for the other without any detectable effect.

The simplest possible statement of equality is that *anything is equal to itself*. In ancient times, Aristotle expressed it as *All A is A*. You could say that *Every dog is a dog*, or *All crooks are crooks*, or even *Truth is truth*.

A statement of this sort conveys no new information to anyone who receives it. It cannot come as a surprise to anyone that something can be used wherever it can be used, nor will such a message tell him any new place that it *can* be used.

Since such a statement gives no new information, it does not appear to be of much use. Early students of logic dubbed it a *tautology* and dismissed it as a faulty technique. Any proof which reduced to a tautology was considered to have proved nothing at all.

It didn't take long for the faults of this approach to be discovered. By the time that Euclid introduced formal proof into plane geometry, the reduction to a tautology was considered to be an essential step of proof.

It's still that way. Any statement of relations which can be reduced to a tautology by any sequence of legitimate operation is automatically proved to be true just by doing those operations.

Take, for instance, the statements the $x + y = 19$ and $x - y = 1$. By performing algebraic manipulations we can combine these two statments into the single statement $2x = 20$, and from this we can get the statement $x = 10$. We can then substitute the value 10 for x in either equation (we'll use the second) and obtain the statement that $y = 9$.

To check these values for x and y, we can replace the x and y in the equation we did not use to find y. This gives us $10 + 9 = 19$. Performing the indicated operation of addition reduces this to the statement $19 = 19$, which is a tautology proving that all our actions were valid.

This example is not the world's most rigorous proof, since it depends upon your acceptance of my claim that $10 + 9$ adds up to 19 and my statements that the algebra is done right in several places. To provide rigor, every step would have to be proved in detail, and every reduction would itself require proof before it could be used.

In 1900, Alfred North Whitehead and Bertrand Russell, who were at the time two of the world's leading mathematicians, set out to do just this for the field of arithmetic. By 1910, they realized that the work was a bit more complicated than they had anticipated, and published the part which they had finished in 10 years. The rest was never published; they were unable to finish it.

Nevertheless, their partial work *Principia Mathematica* marked an epoch in the history of speculative thought, and was one of the first serious attempts to put rigor into math from beginning to end.

The work is far too complex to attempt to describe here: it's mentioned only to emphasize the enormous amount of detail involved in providing true mathematical rigor. For instance, the definition of *number* which we met in Chapter 3 requires 345 pages of preliminary proof in the condensed version of *Principia Mathematica*.

The difference is that Whitehead and Russell provided full rigor before asserting and proving the definition; this volume is intended to unravel ideas, and so accepts the results of their work without independent proof. In consequence, the work you now are reading lacks *rigor*.

This idea of rigor bears emphasis. The word comes from the Latin work meaning stiff, unyielding, unchanging (as in *rigor mortis*), and nothing can be much less yielding than a tautology.

Since mathematical truth must always reduce to a tautology, in mathematics truth and rigor are almost identical. The difference is that rigor involves recounting every detail of proof rather than accepting the word of those who went before.

This means that something can be true without having rigor, and much of what passes for the study of mathematics in most educational institutions is actually the memorization of such mathematical truths which ages of experience have shown to be most useful. Little effort has been expended in the past on providing rigor. Even the so-called New Math makes many assertions without proof in the interest of providing the majority of students with a number of useful tools.

The ideas of proof and rigor normally don't come forth until the student meets *plane geometry*, where the result is often havoc. The student must grasp too many new ideas all at the same time. Geometry deals with pictures rather than numbers, and requires that proofs be thought out rather than reduced to simplified equations according to predefined rules.

Small wonder, then, that many students become convinced that they simply cannot understand the subject.

Other methods of proof besides reduction to tautology exist in the mathematician's arsenal, but none are as foolproof. One which used to be popular was called *reductio ad absurdum* or, translated from Latin, "reducing to absurdity." The proof we used to show that the square root of 2 cannot be a rational number was one of this sort.

To use this technique, you start by assuming that what you want to prove is false. You then show that this assumption leads to a contradiction and so must, by definition, be false. What is not false is true, so if the assumption that your desired proof was false is itself false, the desired proof must be true.

The technique works in many cases. Sometimes, however, it provides a spectactular failure. The first such failure was discovered about 100 years ago, and led (more or less directly) to Einstein's celebrated theories of relativity.

In these cases, the apparent contradiction produced by the *reductio ad absurdum* was not actually a contradiction, but it appeared to be because the workers had not defined their terms properly.

The problem which this brought up in turn, though, was that there is no rigorous method to prove that the contradiction actually exists and is not just the result of a bad definition. For this reason, *reductio ad absurdum* is not presently accepted as a rigorous method of proof, although it is still in wide use as a teaching tool.

For rigor, only those statement which can be reduced to tautologies involving the *primitive postulates* of the universe of discourse can be accepted as mathematically true.

Fortunately, if the universe of discourse is adequately limited and the primitive postulates properly chosen, this is usually possible.

Real-World Truth of Mathematics

We emphasize continually in these pages that mathematics, as such, has nothing to do with the real world: that it describes only an ideal world in which the rules are always followed and in which the actions and rules are greatly simplified.

At the same time, we must always keep in mind that any math which actually has no connection at all with reality cannot be of any practical use to anyone except as a most abstract way to pass time (sort of a super-egghead's game of solitaire).

The apparent contradiction implicit in these two statements is more apparent than real. The ideal world of math can be mapped, more or less accurately, onto the real world of everyday life. When the mapping is extremely accurate, as in the counting of eggs, math appears to be dealing with the real world directly. When it's less accurate, as in computing the position and energy of a single electron in atomic physics, the unreal nature of math quickly becomes apparent.

The usability of math in the real world must obviously depend upon the accuracy of this mapping, which in turn means that we must have some method to assign a *confidence level* to the mapping. That is, we must have some valid grounds for assuming that adding 1 to 19 will get us to 20, rather than 4 or 399, and we must also have a way to find just how much faith to put in the calculated position of an electron when we work with *wave* mechanics.

The measure which we use to determine the accuracy of this tenuous link between math and the real world is called *predictability*. We use math to predict what *will* happen, and we watch the real world to see what *did* happen. We then compare the prediction and the observation.

If they match, good; if not, also good—either way, nothing is proved when we do this one time.

A perfect match one time could have been accidental. So could a total mismatch.

But if they match 10 times out of 10, or 100 times out of 100, you have some reason to believe the mapping is fairly good. By the time it reaches a million matches out of a million tries, you'll be ready to write off a single mismatch as a bad observation rather than bad mapping.

If, out of a thousand tries, not one of the predictions matches, the mapping probably isn't close at all. But what if you hit 750 out of a thousand, and miss 250? This seems to show that the mapping is moderately close, but far enough off to miss one time out of four.

This type of thing is what actually happens at the edges of science; the thing to do is find another mathematical pattern to use and try again. That's what gives us major advances in theories—and keeps making proven knowledge obsolete about the time everyone is ready to accept it as fact.

To sum all this up, you can tell fairly easily when a mathematical operation is true for the real world. If it works, it works (and if a proof by tautology at this point surprises you, you haven't been paying attention). Finding out when they're not so good is a far more difficult matter.

EQUALITY RELATIONS: EQUATIONS

We have seen that the proof of mathematical truth depends upon establishment of an equality relation, specifically upon reducing the statment being proved into a statement that something is equal to itself.

This use alone would suffice to give the equality relation a special prominence among all the possible relations which can be expressed, but equality has more going for it than that.

Arithmetic, for instance, depends entirely upon such equality relations as $1 + 1 = 2$, $4 - 3 = 1$, $2 \times 2 = 4$, and $10/5 = 2$.

Basic algebra also deals primarily with the equality relation; the major difference between this and arithmetic is:

in arithmetic, only constants are involved; in algebra, all kinds of operands may appear.

The uses of the equality relation are so widespread and varied that we give it a special name: we telescope the words *equality relation* into *equation*.

You may have met equations in their guise of being problems for solution rather than seeing them as statements of relation. This is a byproduct of their most common use, but the fact remains that, primarily, an equation is a true statement of the equality relation.

Since equations are such a basic part of mathematics, let's find out in detail how to construct and represent them. We will then be ready to see how mathematical operations are defined, and proceed to some practical applications of math.

Constructing an Equation

We have defined an equation as being a true statement of an equality relation. This implies that every equation is true. If a statement of relation, which claims to be an equation, turns out to be false, that statement cannot be an equation.

The equality relation, you will recall, is one in which the two operands involved are totally interchangeable (within the universe of discourse) without detectable effects. So long as we stay within the universe of *classification by last name*, everyone having the last name of Smith is considered equal to any other Smith without regard to age, sex, race, or other possible distinguishing features—because the universe of discourse for this example is limited to one in which no feature except last name can exist.

We construct an equation primarily to serve as a definition of terms. If, for instance, we know that starting with one egg and adding another gives us two eggs, and adding one more to these two gives us three, we write a pair of equations: The first, $1 + 1 = 2$, tells us that one and one are two, and at the same time defines the operand 2. Similarly, $2 + 1 = 3$ says that two and one are three and simultaneously defines the operand 3.

Note that this is nothing more or less than a trick for recording everyday real-world experiences using the language of mathematics. So far, our equations which define 2 and 3 apply only when we are counting eggs. If we want to count birch trees or grapefruit, we must first do it, then record our experiences as equations. Experience with eggs proves nothing about anything else.

Sooner or later though, we may realize that no matter what we are counting, our written equations come out the same.

That's the point at which mathematics suddenly becomes useful. If we write the same equations every time we record the results of a count, it's a small extrapolation to guess that the equations might apply to the act of counting, rather than to the eggs or trees or fruit.

While the extrapolation is small, the logical step which it involves is unimaginably large. No matter how many times we write the equations, it's still only a probability that counting will work as we expect it to, if it is applied to something nobody ever tried it on before. The probability is very great, true, but it's still not a certainty.

Before we lose sight of just what we did when we wrote these equations, let's take another look. All we did, in fact, was to give a name to the result of an action, and record that name together with the action so that we could refer to it in the future. After doing this a number of times, we recognized that we were using the same name repeatedly.

Let's look in a little more detail at what we did. We took the action of adding one and one and wrote it as $1 + 1$, then named the result 2 and wrote that down also, conncting the first and second operands with an *equal* sign: $1 + 1 = 2$. This recorded the action and the name for posterity, which in turn lets us use 2 as a constant and do it all over again with $2 + 1 = 3$.

What we are really doing, then, when we write an equation of this sort is giving a name to the result of an action. It would be more obvious if we wrote $2 = 1 + 1$ and $3 = 2 + 1$, instead of the other way around; the equations would be equally valid, since either operand of an equation can substitute for the other. If we substitute on both sides, we turn the equation around, if we substitute on only one side, we get a tautology.

Since this type of equation simply assigns a name or a symbol to an operand, no question of proving its truth can arise. The equation is true just because you say it is. That is, if you write the equation $x = 3$, you are, in effect, saying *I have given the name x to the value 3*, and since you are free to give any name you like to any value (unless you have already used the name for some other value which will come back into the acivities again), no question of truth is possible.

When you consider two or more equations at the same time, the situation changes. You now have a system of equations, which can be consistent only if every equation in it really is an equation, and none of them contradict any of the others.

Most practical applications of equations involve systems of equations, so let's look a little closer at them.

Systems of Equations

A system of equations is a set of two or more equations considered at the same time (that is, within the same universe and simultaneously). All of the equations must be true, and none must contradict any of the others for the system to be consistent. A system of equations which is not consistent has no usable meaning.

For instance, you could write two equations: $x = 2$ and $x = 3$. This forms a system, since more than one equation is involved. Each equation, taken by itself, is true. Together, though, each contradicts the other. The only way this system could become consistent would be to consider $2 = 3$ a true statement, and that relation is false for any defined number system used in mathematics.

The result is that we have an inconsistent system. In such a system, we can draw no conclusions. In particular, we cannot use the inconsistency to prove that either equation is false. All that we can legitimately say about such a system is that within the system, if one is true the other cannot be, and so the system (but *not* either equation) contradicts itself.

This would be true even if we were looking at a system of 1000 equations made inconsistent by just one of the thousand. Removing that equation from the system would leave us a consistent system of 999 equations, but this does not prove the one to be false. All it proves is that it is not consistent with the system.

Whenever we come up with an inconsistent system of equations we have methematical nonsense. All the grammar is correct, each separate equation by itself is impeccable, but the system as a system has no assignable meaning.

It's like *Alice In Wonderland* (which, by the way, was written by one of the leading mathematicians of his time), in that every little part seems plausible enough—but when you look at the whole thing, it just doesn't make sense.

Just as the equation is a sort of shorthand for recording experience or actions, we have available to us a shorthand we

0+1=1	0+2=2	0+3=3	0+4=4
1+1=2	1+2=3	1+3=4	1+4=5
2+1=3	2+2=4	2+3=5	2+4=6
3+1=4	3+2=5	3+3=6	3+4=7
4+1=5	4+2=6	4+3=7	4+4=8
5+1=6	5+2=7	5+3=8	5+4=9
6+1=7	6+2=8	6+3=9	6+4=0
7+1=8	7+2=9	7+3=0	7+4=1
8+1=9	8+2=0	8+3=1	8+4=2
9+1=0	9+2=1	9+3=2	9+4=3
0+1=0	0+2=0	0+3=0	0+4=0
1+1=0	1+2=0	1+3=0	1+4=0
2+1=0	2+2=0	2+3=0	2+4=0
3+1=0	3+2=0	3+3=0	3+4=0
4+1=0	4+2=0	4+3=0	4+4=0
5+1=0	5+2=0	5+3=0	5+4=0
6+1=0	6+2=0	6+3=0	6+4=1
7+1=0	7+2=0	7+3=1	7+4=1
8+1=0	8+2=1	8+3=1	8+4=1
9+1=1	9+2=1	9+3=1	9+4=1

Fig. 4-4. The equations shown here are only a part of the system of 200 separate equations that define decimal addition. Those at the top define the sum digit, while those at the bottom define the **carry** digit.

can use to record systems of equations. Sometimes it's called a *structure*, other times it's known as an *array* or *matrix*, but in everyday use we call it a *table*.

We've made use of tables before, and they're common in the real world. The only new aspect of them we're meeting here is their application to show systems of equations.

One of the most often used systems of equations is the one which defines what we call *addition*. We'll see in Chapter 5 how it is built from the primitive postulates of *zero* and *the successor function*; right now we're using it only to illustrate how a table can show a system of equations.

Figure 4-4 shows the first few equations of the system. For use with the decimal number system, this system of equations contains 100 equations defining the *sum* digit and another 100 equations to define the *carry* digit.

That same system of equations, expressed as a table, is shown in Fig. 4-5. This version includes all 200 equations of the basic decimal set; Fig. 4-4 shows just a few of them.

You can see how much easier the table is to use compared to the separately listed system. To find the sum digit to be substituted for any pair of digits which are being added, simply read across the row which begins with one digit of the pair until you reach the column headed by the other digit of the pair, and use the digit which appears in the upper portion of the intersection. For a *carry* digit, use the digit in the lower portion of the intersection. It's that simple.

For instance, to recover the information originally given by the system equations $3 + 2 = 5$ for the sum digit and $3 + 2 = 0$ for the carry digit, we read down past rows **0**, **1**, and **2** until we get to the fourth row (headed **3**). Then we read across, past columns **0** and **1**, so that we read the intersection of row **3** and column **2**. There we find the sum digit *5* and the carry digit *0*.

The equation is still there, but it has been compressed so that one takes care of all 20 equations for which 3 is the first operand, and one 2 similarly serves for all 20 equations in which a 2 follows the symbol + .

The systems of equations which define what we know as addition, subtraction, multiplication, and division form the basis of arithmetic, and we will look at them in much more detail in the next chapter. They are not, however, the only kinds of equation systems which are important to us.

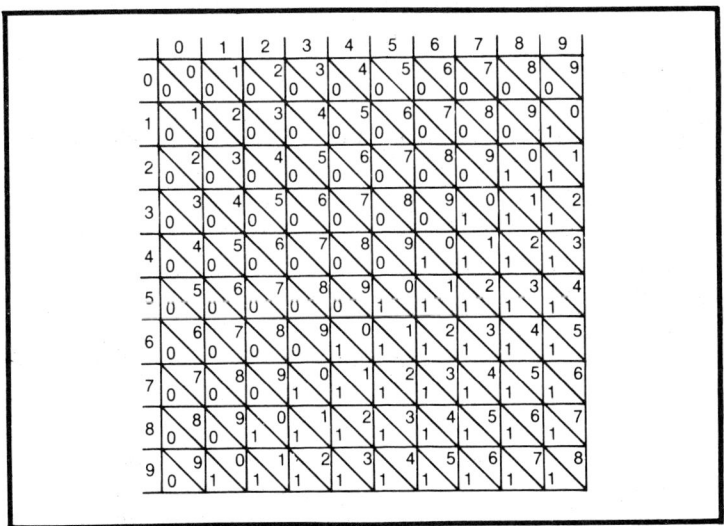

Fig. 4-5. This table presents the entire system of 20 equations defining decimal addition. Sum digits are in upper right part of row/column intersections, with carry digits in lower left. This is the decimal addition table that defines the operation. We'll meet it again in the next chapter.

When we were examining the idea of mathematical truth, we met another kind of equation system. There, we used a pair of equations: $x + y = 19$ and $x - y = 1$.

Since more than one equation is involved, we have a system. What makes this system different from the other systems we've looked at so far is that it deals with two different variables.

The system can be consistent only if both equations are true at the same time. As it happens, only one value for x and one value for y will make the system consistent, although either equation taken by itself has an unlimited number of possible answers.

In algebra, this is called a system of *simultaneous* equations, and the specific system we're discussing would be called *a pair of simultaneous linear equations in two unknowns*.

Actually, as we have seen, any system of equations requires for consistency that all its equations be true simultaneously, so the name given in algebra is not quite as defining as it might be, but we cannot change decades of tradition.

The study of simultaneous equations forms a major part of classic algebra and provides may of the practical engineering applications for mathematics. These systems, too, may be expressed in terms of tables or matrixes, and one entire branch of math known as *matrix algebra* deals with operations upon matrixes, for the purpose of solving systems of simultaneous equations.

Still another kind of system of equations forms the basis of the differential calculus and the integral calculus, which despite their imposing names and reputations for complexity are little if any more difficult than algebra, and certainly are simpler than the geometries.

The systems of equations employed here deal with what happens to *dependent variables*, as their controlling *independent* variables are permitted to vary over some range. The systems of equations define the relations between the *independent* and *dependent* variables. The study of the changes which occur is what we call *differential calculus*. *Integral calculus* is an inverse study, which attempts to determine the system of equations involved from examination of the changes which result.

Chapter 5

Recipes and Rules

The verbs of mathemaics are those things called operators which we have been tossing about since Chapter 1. In general, an operator is anything which connects two operands and calls for some operation to be performed. A relation, by contrast, connects two operands to make a statement.

As many kinds of operators are possible as there are possible kinds of relations. In both cases, the only limit is your imagination. However, if an operator is going to be of any use to anyone it must call for some defined operation to be performed, so if you're going to invent an operator, you must also invent an operation for it to perform. This, in turn, involves definition of the *result* of the operation for each allowable pair of operands.

This definition of results is done by equations, using at least one equation for each pair of operands, so that *defining an operation* is another way of saying *constructing a system of equations*. This system, which must be consistent if it is to be useful, is normally represented in tabular form, which tends to obscure the fact that each useful operator is defined by a consistent system of equations.

Almost all of the math we learn in school boils down to learning the use of the tables for a rather small set of operations. Since you probably have some familiarity with this kind of math—although not necessarily under the names we're using for it right now—we'll start our look at operators with the four basic operations of arithmetic.

THE VERBS OF ARITHMETIC

Basic arithmetic, the kind taught in elementary school, involves four different operators. Sometimes we call them *plus, minus, times,* and *divided by,* and sometimes we use the symbols +, −, ×, and / for them. We may also call them by name as *addition, subtraction, multiplication,* and *division.*

Each of these operations involves its own system of equations, which can be represented as a table. Most of us have heard of the multiplication table, which is just the tabular presentation of the 200 equations which define the product and carry digits for the 10 decimal digits taken by pairs.

It would be just as correct to talk of the addition table or the subtraction table, but the traditional way of teaching arithmetic didn't involve memorizing these tables quite so directly.

Actually, we all did have to learn these tables, but we did it so early in our educations that we didn't know what the word *table* meant in math. We simply learned one and one are two, two and two are four, and so forth. That is, instead of learning the additional table in the first and second grades, we memorized the entire system of equations for it. Small wonder it took two full years!

Then the teachers spent more time convincing us that two and three were just the same as three and two, after we had spent so much time learning them as two separate equations.

Hopefully, as we trace through the tabular approach to operators and operations in this chapter, we will see the underlying patterns which permeate the whole subject, and thus learn to extrapolate it for our own specific needs as they arise.

Addition Tables

In Chapter 4, when we made the acquaintance of systems of equations, we saw that 200 equations were required to define *addition* for the 10 decimal digits. Of these, 100 specify the *sum* digit and the other 100 specify the *carry.*

In this system, each of the 100 *carry* equations defines the carry digit as either *0* or *1*; no other digit is permissible as the carry. To see why this is so, and to develop the 200 equations, let's take a step backward and look at simple counting one more time.

For this discussion, let's define three identical sets of digits, each of which contains 10 members. These members are the decimal digits *0* through *9*. Each member of the first set, which we'll call the *units* set, is worth one count. Members of the second set, which we'll call *tens*, are worth 10 counts each. The third set, which we'll call *hundreds*, has members with a value of 10 *tens*, or 100 units each.

We'll form our count by taking one member from each of the three sets. At the start, before we have counted anything, we take *0 from each set, which gives us a total tally of 000* counts.

As we count, we'll follow the basic rule that we take the next member from the *units* set for each item tallied, and put back the one we had. When we use the *9* from any set, the next item will cause us to replace the *9* with the *0* in that set and at the same time take the next member from the next most significant set.

On the first count, we replace the *units* 0 with the *1* and get a tally of *001*. This continues until we have a tally of *009*; the next item uses the *carry* to step to 010 and so start the *units* action all over again.

Within the universe of discourse as we have defined it, you can count up to *999*. The next time rolls the tally over to *0*, because we do not have a fourth set of digits.

Note that we are looking at *counting*, not addition. In order to record our results for the 999 counts possible with only three sets of digits, we would have to have a system of 999 equations—one for each possible outcome.

However, as we perform our 999 possible tallies, a pattern emerges. In fact, we used this pattern to describe how the process works. Every time we pass from *9* to *0* in one set, we generate a *carry* digit which impersonates a count in the next higher set.

Now that we have recognized the pattern, we no longer need restrict our universe to three sets of digits. One set will do, if we can copy it as needed. That's what we mean by the phrase *positional notation*.

When we use this technique, we start with just one set of digits or *one place* figures. As soon as the passage from 9 to 0 generates a carry digit, we make a new copy in the next higher

SUM	CARRY
0+1=1	0+1=0
1+1=2	1+1=0
2+1=3	2+1=0
3+1=4	3+1=0
4+1=5	4+1=0
5+1=6	5+1=0
6+1=7	6+1=0
7+1=8	7+1=0
8+1=9	8+1=0
9+1=10	9+1=1

Fig. 5-1. These 20 equations form a system which defines the operation of counting.

place and fill it with the carry digit, which initially is a *1*. This gives us the transitions fom *9* to *10*, *99* to *100*, *999* to *1000*, and so on. Our digit sets are now unlimited.

We can now record what happens for one complete counting cycle through the set of digits. It takes 10 equations to represent the 10 possible *sum* or replacement digits, and 10 more to represent the *carry* digits. All of the carry digits are *0* except for the last one, which *generates the carry*; this lets us add the carry into the next place every time, thus expressing our rule as a set of equations.

Figure 5-1 shows this system of 20 equations which define the counting process when decimal numbers are employed.

Notice that the number of equations required depends upon the number of members in the digit set. When we use 8 rather than 10 (octal rather than decimal numbers), we will have eight digits (0 through 7) and only 16 equations in our system. If we use binary (two digits, *0* and *1*), we will have only four equations to deal with.

From counting to addition is a very short step, both logically and intuitively. Counting is, in fact, the same thing as adding one to the previous tally.

If we count by two, we add two instead of one at each tally, and so forth. We can extend this idea to generate a table or define a system of 200 equations, which defines the process through *counting by nines*, and call this system a definition of the *addition* operator.

If we do this and begin to record the equations, the pattern becomes apparent. The difference between the list of 10 *sum* equations for counting by ones, and that for counting by twos, is that each sum in the twos list is one unit greater than the corresponding etry in the *ones* list.

This one-unit offset continues through the system for each succeeding list, giving us the system tabulated in Fig. 5-2. In this pair of tables, the *sum* table shows the digit to use in the current position of the result, and the *carry* table shows the digit to be added to the next most significant position. Notice that the largest carry which can be generated is that which results when we add 9 to 9; since the result is 18, the carry is 1. No larger carry can be generated.

The pattern of a one-unit offset sows up in the *sum* table as the diagonal placement of the sum digits. Notice how the 9 sum forms a diagonal from lower left to upper right of the table. The *carry* table shows a similar division between 0 and 1, since the carry changes from 0 to 1 when the sum digit passes from 9 to 0.

We can use this same approach to generate a pair of tables defining addition for octal numbers, as shown in Fig. 5-3. The

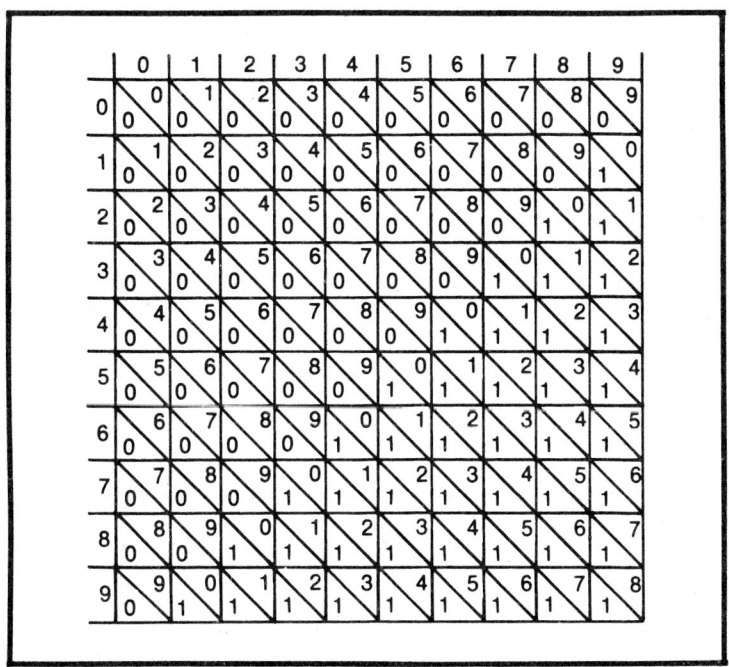

Fig. 5-2. The decimal addition table is developed from the counting equations, by extending them and repeating them to cover the whole set of numerals.

105

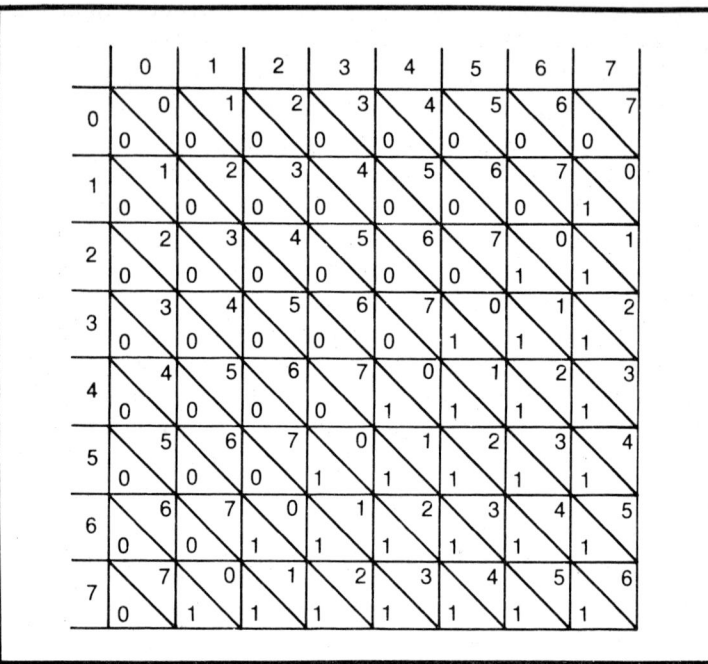

Fig. 5-3. The addition table for the octal number system is just like that for decimal numbers except that digits **8** and **9** are missing, which makes the carry digit show up sooner and leads to unfamiliar representations for larger numbers.

only difference is that the carry appears sooner, since we do not have the 8 and 9 digits available.

When we get down to the addition table for binary numbers (Fig. 5-4), we have the simplest such table possible. The simplicity of this table is what made electronic computers possible.

All of these addition tables, though, follow the same basic pattern, and it's based on the ideas of counting. When we get

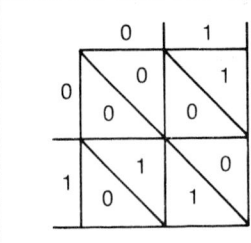

Fig. 5-4. When you get to the binary number system, the table for addition is so simple it can be built into a machine. Carry exists only when adding **1** and **1**.

right down to it, addition is nothing more than a shortcut for counting, so that we don't have to count up to the answer every time. And while that may sound almost trivial, it's not—nobody could work problems in large numbers without the shortcuts made possible by the addition tables.

One important property of the addition operation, which shows up in the tables, is its symmetry. If you read directly across the 7 row of the sum table in Fig. 5-3, you'll see that it contains the same digits in the same sequence as does the 7 column of the same table when read straight down.

Because of this symmetry in the tables, the operation of addition is said to be *commutative*. That's how mathematicians say that when you want to add 2 and 3, it makes no difference which you write first in the operation. The expression 2 + 3 is exactly the same as 3 + 2 because either is equal to 5.

Any operation which behaves like this is called *commutative*. Many, but not all, of the important operations in mathematics are commutative. One which is *not* is subtraction.

Subtraction Tables

The operation of subtraction, indicated by the minus — operator symbol, is also related to counting. Unlike addition, though, the count goes down rather than up. Since this is a *taking away* rather than an *adding to*, it makes a difference *which* you take from *what*.

The tables which provide the remainder can be developed just as we did for addition, with one major difference.

Since subtraction is no commutative—that is, the result of 5 − 3 is different from that of 3 − 5—the rows and columns of the table will not be interchangeable. That is, if the rows correspond to the first operand or *minuend*, the columns must then correspond to the second operand or *subtrahend*. The intersections will then correspond to the *difference* for one table, and to the *borrow* for the other.

Since subtraction is a shortcut for counting down rather than up, almost everything about it is reversed as compared to addition. Addition begins at the units digit and carries over to the more significant digits. Subtraction starts at the most significant digit of the larger operand and borrows counts as necessary from the less significant digits.

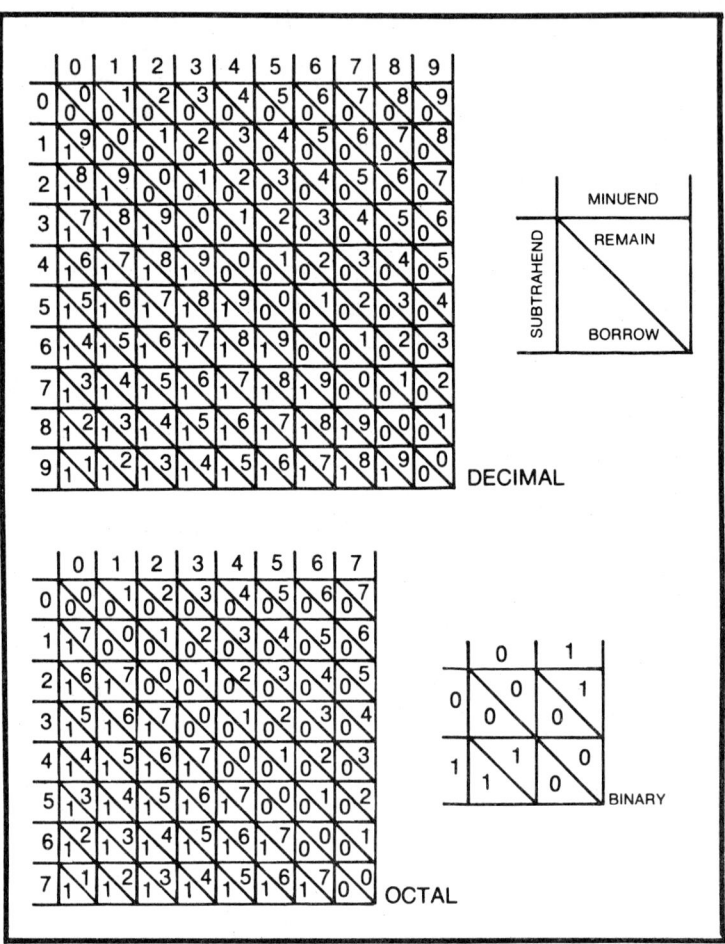

Fig. 5-5. Subtraction table for the three number systems is similar to the addition tables, but use is more demanding. Columns represent **minuend** digits (being taken away from) and rows are **subtrahend** digits (being taken away). **Carry** is replaced by **borrow** when subtracting larger digit from smaller.

Keeping these differences in mind, you can compare the tables in Fig. 5-5 for decimal, octal, and binary subtraction with those for addition (Fig. 5-2 through 5-4) and see the relation, the similarities, and the points of difference.

Multiplication Tables

The first operation table most of us ever met was most likely the multiplication table. For many students, it was bad news, since it had to be memorized to be of any use—but then, so must all the other operation tables of mathematics.

Actually, multiplication (symbolized by × and indicated in many computer languages by *) is nothing more than a shortcut for addition, just as addition itself is a shortcut for counting. Just as the addition table records the results of repeated counts, so the multiplication table records the results of repeated additions.

We use positional notation with the multiplication table, just as we do with the addition table. The operands are looked up by row and column to find the product (which corresponds to the sum of the addition table), and again in another table to find the carry.

One major difference between multiplication and addition is that in multiplication, the carry digit is not limited to 0 or 1, but can be any digit of the number system (except the last). Exclusion of the last digit means that binary multiplication cannot generate a carry other than 0, octal multiplication cannot generate a carry of 7, and decimal multiplication cannot generate a carry digit of 9.

Like addition, multiplication is commutative. Operands can be interchanged with no effect on either product or carry. Figure 5-6 shows binary, octal, and decimal multiplication tables.

Division Tables

The fourth basic operation of arithmetic division, bears the same general relationship to subtraction that multiplication does to addition—but there's an essential difference as well. Multiplication is a shortcut for repeated addition, and division is a shortcut for repeated subtraction. Addition and multiplication are commutative; neither subtraction nor division has this property.

But subtraction is an exact inverse of addition. That is, if $2 + 3 = 5$, then $5 - 3 = 2$.

With division and multiplication, if we stick to the set of integers, it's not quite so exact. So long as we deal only with numbers which are exact multiples, division and multiplication are true inverses one to the other.

But if we pick just any two numbers at random, we may find that multiplication does not exactly reverse the division act. For instance, if we try to divide 9 by 7 (within the set of

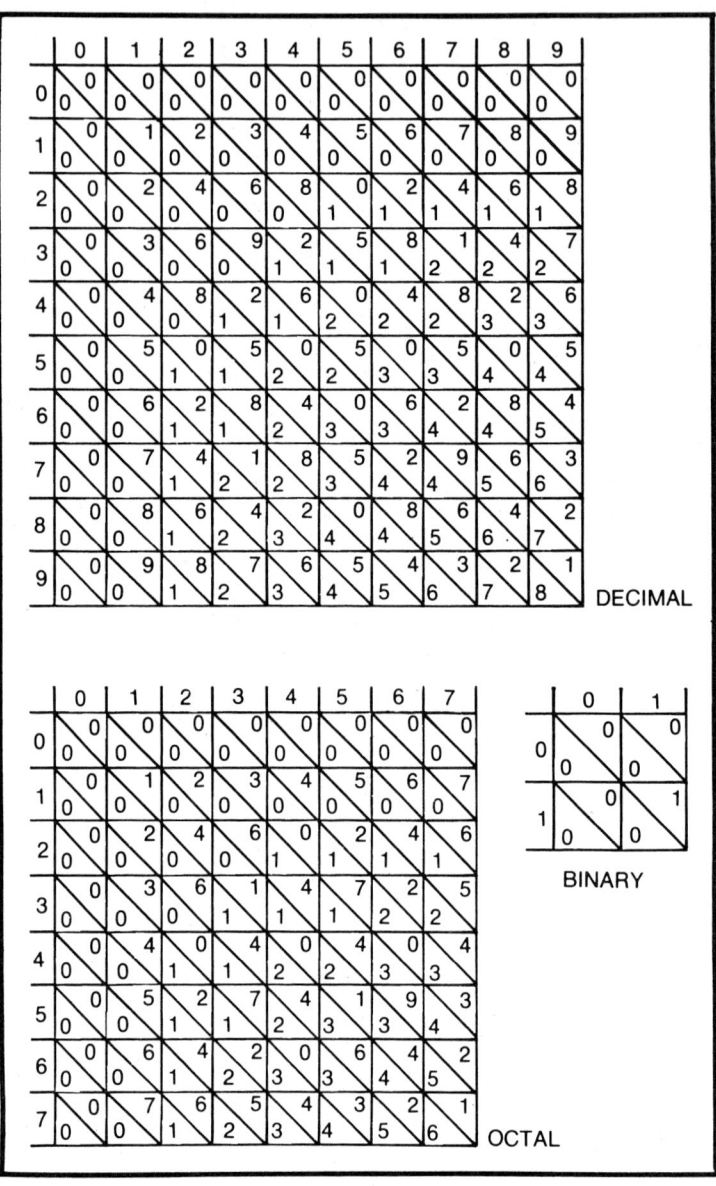

Fig. 5-6. Here are the multiplication tables for the three number systems. Notice that carry cannot be produced for binary, and zero times anything remains zero in all three systems.

integers) we get a result of 1 and a *remainder*, or uncompleted borrow, of 2. When we multiply 7 times 1 we get 7, not 9. To get the total of 9, we have to add in that remainder as well.

Thus, while division is *an* inverse of multiplication, it's not an exact inverse over the set of integers. The reason for this is that integer division is a many-to-one mapping, and we saw earlier that such a mapping cannot be reversed. Thus *integer division has no true inverse operation*.

When we move from the set of integers to the set of real numbers, the problem disappears. Over the real numbers, division is the exact inverse of multiplication—and that's one of the reasons why it was necessary to define the real numbers in the first place.

The division table, like the table of all integers, has no limit and so we cannot show it all. That portion of it shown in Fig. 5-7 shows the basic pattern for integer division. Note that division by zero is undefined, and any number divided by 1 is equal to itself.

Within each column of the table, you can see the pattern involved in counting to the corresponding base. Each time the *remainder* part of the table becomes *0*, the *quotient* entry becomes one unit larger. This is, actually, a counting up of the remainder to the base established by the divisor.

For everyday use, we don't try to use a table for division. We isolate the pattern which shows up in the table, restate it as a procedure or *algorithm* for the operation, then perform the procedure.

Most often, we use the real-number form of division rather than the integer form. This carries the quotient out to some arbitrary number of decimal places.

Another often used way to express the result of division is in the form of fractions. Any fraction is a rational number, because it is shown as the ratio of one integer to another (even though it may not necessarily be reduced to lowest terms).

When doing *real* division, if the number of decimal places is limited to some arbitrary number, the result is also a rational number. The number of places you choose to use will establish some power of ten (in decimal numbers) as the denominator of a fraction, and the quotient forms the numerator of that fraction.

For instance, let's take the figure often used to represent pi. Pi is the ratio of the circumference of a circle to its diameter, and the actual number involved is one of the real numbers which cannot ever be exactly expressed in decimal digits. The value has been computed to several thousand

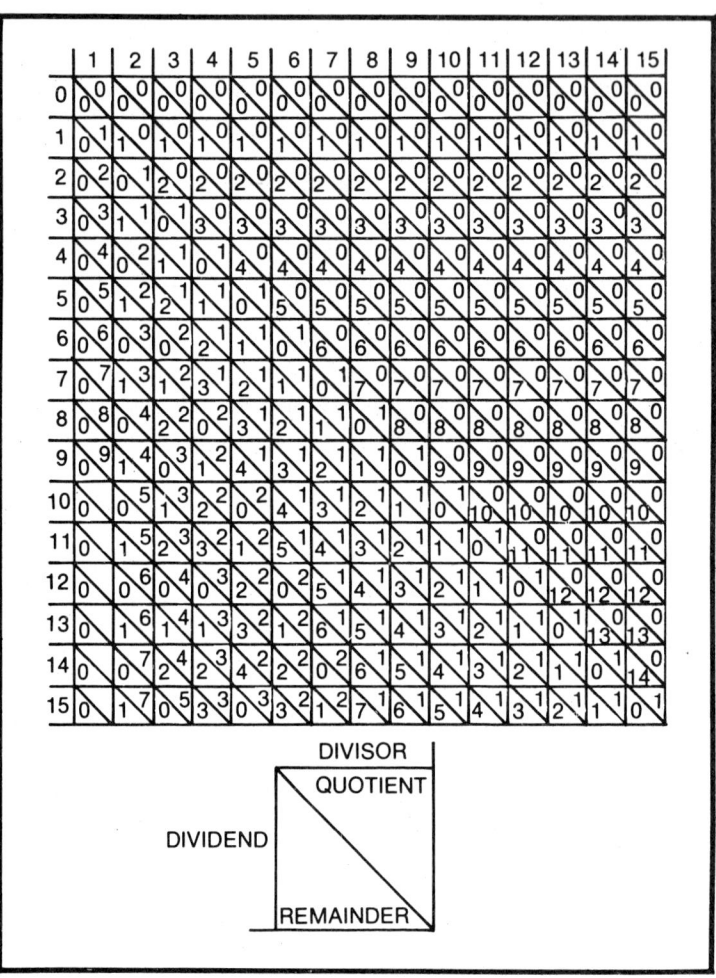

Fig. 5-7. This is a part of the decimal division table. It's unlimited in both directions, so we cannot show it all. Like subtraction, its use follows strict rules. Columns are divisor digits (or numbers, in this case), and rows are dividends. Intersections provide quotient and remainder.

places, but engineers ordinarily use the value *3.14159* when they need to be very accurate, and *3.14* as a rough value.

The number 3.14159 is actually the fraction 314159/100000, which is a rational number. What's more, it's in its lowest terms, since *314159* is a prime number having no factors but itself and *1*.

About the only place where the integer form of division has application in everyday life is in the making of change, and the way most people count out change tends to hide the fact that any kind of division is involved.

For instance, if you buy something which costs 21 cents, and give the sales clerk a dollar bill, you should be given 79 cents change. You'll probably get three quarters and four pennies, counted out as "...22, 23, 24, 25, 50, 75, one dollar."

You can, however, see the integer division process at work if you divide the 79-cent change requirement by the 25-cent value of a quarter, getting a quotient of 3 and remainder of 4. This means 3 quarters are necessary, and there's 4 cents to go.

Clerks are taught to count up from the change drawer, starting with the price of the purchase and adding pennies until the count reaches a multiple of 5. If the multiple of 5 first reached is not a multiple of 10 or 25, nickels may be counted in to reach a multiple of 10 or 25. The dimes, quarters, and half-dollars are counted in as required, and the same sequence is used to go up through the paper money as far as necessary.

This method is used because most merchants have learned (the hard way) to trust counting more than mental mathematics. The principles, however, remain valid.

MORE EXOTIC IDEAS

Now that we have made the acquaintance of the four basic operators of arithmetic and have seen how each of the four can be defined by a system of equations or a set of tables, let's look a little more closely at some of the operators encountered less frequently.

Powers and Roots

We saw that the act of counting could be summarized in a set of tables to produce the operation of addition, and that addition could in turn be summarized in another set of tables to produce multiplication. Can the process be continued?

Yes, it can. The next step is called by several names, but in general it bears the same relation to multiplication that multiplication does to addition, or addition to counting.

While we do not know for certain just when addition or multiplication was invented, we do know when the operation of raising to a power was formalized; Descartes, the inventor of analytic geometry and Cartesian coordinates, is credited with establishing it as an operation separate from that of multiplication.

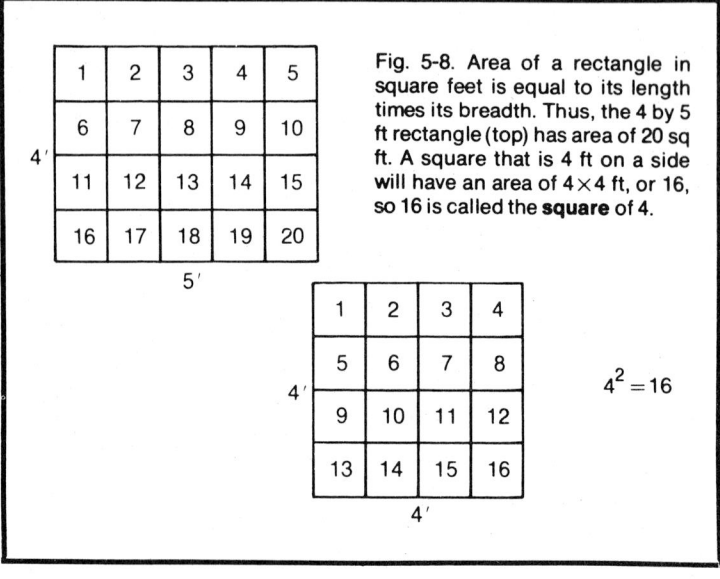

Fig. 5-8. Area of a rectangle in square feet is equal to its length times its breadth. Thus, the 4 by 5 ft rectangle (top) has area of 20 sq ft. A square that is 4 ft on a side will have an area of 4×4 ft, or 16, so 16 is called the **square** of 4.

Certain special cases of the operation have been known for centuries before Descartes established the symbolism, and the names used for these special cases are still part of mathematics. They are the *square* and the *cube* of numbers; these names come from the original applications by the Greek geometers.

The geometers were interested in measuring the earth. That's what geometry means: earth measurement. To do this, they had to solve problems involving area and volume as well as length. The area of a rectangle is found by multiplying its length and width, so the area of a square is equal to the length of any side multiplied by itself (Fig. 5-8). Because of this, they called the product $A \times A$ the square of A.

Similarly, the volume of a rectangular box is figured by multiplying together its length, width, and depth (Fig. 5-9), so the volume of a perfect cube would be equal to the length of any edge multiplied by itself twice. Thus, they called the product $A \times A \times A$ the cube of A.

Since the ancient geometers dealt only with the three dimensions of length, width, and depth, they went no higher than three factors and didn't invent a name for the product $A \times A \times A \times A$ or any higher orders of the operation.

Not until Descartes defined the operation of raising to a power did repeated multiplication obtain a unique symbol.

What he did was to tally the number of times the number appeared as a factor, and append this count as a superscript above and to the right of the number itself. Thus the square of A, written Descartes' way, would be A^2, and the cube of A is A^3.

We still call these quantities the *square* and the *cube* of A, but A^4 is known merely as *A to the fourth power* or *the fourth power of A*.

The two operands involved in this operation are called the *root* and the *exponent*, respectively. That is, in the operation A^7, the root is A and the exponent is 7. The operator is indicated by writing the exponent as a superscript. Some computer languages indicate this operation by the symbol ** and others use a vertical arrow betwen the operands.

Unlike the less exotic operators, the operation of raising to a power does not lend itself to tabular presentation. Like division, the table is unlimited; the operation is defined by the procedure used rather than by a table.

Like most operations, raising to a power implies an inverse which we call *extraction of the root*. To square a number, we multiply it by itself. To extract the square root, we

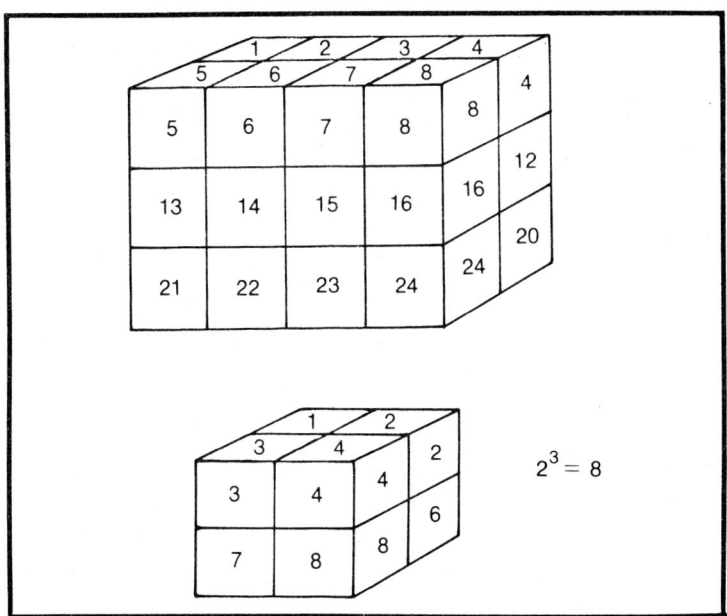

Fig. 5-9. Volume of a solid is figured like area of a plane, length times breadth times depth. Thus 2 ft. cube has volume of 8 cu ft, and 8 is called **cube** of 2.

115

determine what number, when squared, will yield the number with which we started.

This is a relatively simple process despite the fact that almost all square roots are irrational real numbers which can never be exactly represented. Newton discovered the simplest rule for extracting the square root, and we'll go through it in just a moment. First, let's see why most square roots are irrational.

Let's start with something that may strike you as being a backward approach; let's list the squares of the first 10 integers—1, 2, 3, 4, 5, 6, 7, 8, 9, and 10.

The square (and, for that matter, any other power) or 1 is 1 itself. That of 2 is 4; of 3, 9; of 4, 16. The other six squares, in order, are 36, 49, 64, 81, and 100.

That means that of the 100 integers in the range 1 through 100, only 10 of them have integers for square roots. The other 90 may be rational, or they may be irrational real numbers.

We have already seen that the square root of 2 is not rational. That for 3 must lie between 1 and 2, and can also be proved to be irrational.

The square roots of the integers 5 through 8 must be between 2 and 3. The square root of 8 is particularly interesting, since 8 is the product of 2×4. Its square root, presumably, would be the product of the square root of 2 times the squareroot of 4, or twice the square root of 2. Thus it too is irrational.

We can bracket the integers between which the various square roots lie in the range we're looking at.

When we do, we'll find that no square roots of integers are found in the range between 0 and 1. Only two are in the range between 1 and 2. Four are in the range between 2 and 3. Six occur between 3 and 4. Eight are in the range between 4 and 5, 10 between 5 and 6, 12 between 6 and 7, and so forth.

As we go through the integers, we find that each time we move one number higher, we find two more square roots than in the previous step. Another way of saying this same thing is that *the number of square roots of integers which lie between any two adjacent integers is equal to two times the smaller of the two integers.*

This rule can be proved rigorously; it stems from the relationships involved in addition, multiplication, and squaring of numbers.

If we keep going higher and higher in the range of integers, we will have more and more square roots between adjacent integers. The number increases without limit.

This does not, in itself, prove that the roots must be irrational. Not all square roots are irrational. The square root of 9/16, for example, is 3/4—a rational number yet not an integer.

It does, however, indicate that some square roots are irrational, and gives us cause to suspect that as we continue into ever larger numbers, the percentage of irrational roots will continue to increase just as the total number of roots increases. While this suspicion can be proved, the strict proof is too complicated to go into here.

Now let's see how Newton's technique lets us extract as close an approximation as we like to the square root of any number.

Since we know that the square root of 2 is irrational, let's try to find it for our example.

The first step of the process is to take a guess at the value. Since we know that 1×1 is 1 and 2×2 is 4, the square root of 2 must be somewhere between 1 and 2. Let's split the difference and use 1.5.

Next, divide the number by the guessed value: 2/1.5 gives us 4/3 or 1.33333333. We can carry the 3s out as far as we like since it's a repeating fraction. Let's stop with 1.33333.

Now comes the key step. We add our trial value, 1.5, to the quotient just obtained, 1.33333, then divide by 2 to get a new trial value. This gives us $1.5 + 1.33333 = 2.83333$, which divided by 2 is 1.41667.

That's the result of our first round. We can check its accuracy by squaring it; we get 2.0069428889 as the result. Since this is a little too large, we can improve accuracy by going around again. This time, we use our first result as the new trial value.

When we divide 2 by 1.41667, using an 8-place pocket calculator, we get 1.4117613. When we add this to the trial value and divide by 2, we get a new trial value of 1.4142156. This value, squared, equals 2.0000057. We are much closer, but not there yet.

We can keep going around the loop as long as we like. Each time around, we get closer to the unreachable actual root. In most real-world applications, the value used for the square root of 2 is 1.414 (which squares out to 1.99396), and Newton's method reached this value in two tries.

The technique is simple, but the arithmetic is painful unless you have a calculator. First, guess a trial value, then divide it into the number whose root is to be extracted. Split the difference by adding the guess to the quotient and dividing by two to get a new trial value. Check it by squaring it, and if you're not close enough, go back and do it all over again using the new trial value in place of the old one.

Even if the guess is not very good, the technique will get you to an accurate result very rapidly. For instance, we know that the square root of 3 must also be between 1 and 2, since the square root of 4 is 2—but let's try to find that square root, and use 4 as our first trial value.

First we divide 3 by 4 to get 0.75. Adding 3 and 0.75 gives us 3.75, and dividing this by 2 gives us 1.875. When we square our first trial value, 1.875, to test it, we get 3.515625. Even though our guess was deliberately far too large, the technique brought us fairly close on the first try.

Now we do it again: 3 divided by 1.875 gives us 1.600. Adding the two values gives us 3.475, and dividing by 2 makes our second trial value 1.7375. This value, squared, gives 3.0189062. We're much closer.

A third time around: 3 divided by 1.7375 gives 1.7266187 according to the 8-place calculator. This produces a split difference third trial value of 1.7320593, which when squared is 3.0000294. We may be as close as we need to be, but let's go around one more time to make sure. By this stage, incidentally, the calculator is introducing roundoff errors, so we may not be able to get any significant improvement.

For our forth and final try, 3 divided by 1.7320593 gives 1.7320423. Summing and halving gives 1.7320508. When we square this, we read 2.9999999. That's as close as we are likely to get.

Note that Newton's method, as explained, works only for square roots. A similar direct technique for extraction of cube roots, or those of any higher power, is shown in Fig. 5-10. Like the square-root technique, it keeps going until you have the accuracy and precision you want.

Note also that any number has not one but *two* square roots. Since the square of a negative number is the same as the square of a positive number, the square root of a number can have either sign—which means that every square has two roots.

Similarly, every number has three cube roots, four 4th roots, and so forth. The techniques mentioned here find only some of the roots of any number; they do not find all the roots.

Logarithms and Their Inverses

If we have to go round and round the loop procedure of Newton's method and its relatives to find arbitrary roots, and have no tables to guide us in raising an operand to whatever power strikes our fancy, what practical use can be made of powers and roots?

Actually, quite a bit. Instead of the tricks and tables, we have another pair of operations available for use. These operations help us get useful results from problems dealing with powers and roots. To see what the operations are and how they work, let's take what might at first seem to be a giant step backward, and look at the binary number system.

Binary deals with only the digits 0 and 1, so its base must be two, the total of the number of digits used in the system. This means that the digit positions represent powers of two.

---

to extract: $\sqrt[3]{27}$

$X_1 = 4$ (GUESSED)

$X_2 = \left[\dfrac{27}{4 \times 4 \times 4} + (3 - 1) \right] \times \dfrac{4}{3} =$

$X_3 = \left[\dfrac{27}{2 \times X_2 \times X_2} + (3 - 1) \right] \times \dfrac{X_2}{3} =$

$X_4 = \left[\dfrac{27}{X_3 \times X_3 \times X_3} + (3 - 1) \right] \times \dfrac{X_3}{3} = 2.9999999$

to extract: $\sqrt[R]{Y}$

$X_{N+1} = \left[\dfrac{Y}{(X_N)^R} + (R - 1) \right] \times \dfrac{X_N}{R}$

Fig. 5-10. This is how to find any root of any number by iteration. Cube (third) root of 27 is found in 4 steps as 2.9999999; actually, it's 3. Arithmetic gets tedious in this method; it's recommended only if you use a pocket calculator to do the calculating.

N	2^N	N	2^N
0	1	10	1024
1	2	11	2048
2	4	12	4096
3	8	13	8192
4	16	14	16384
5	32	15	32768
6	5	16	65536
7	128	17	131072
8	256	18	262144
9	512	19	524288
		20	1048576

Fig. 5-11. This table shows the first 20 powers of 2. Each is just twice the preceding value. Powers of 2 are widely used in computing.

The first position to the left of the units position represents the first power of two, which is (in decimal) 2. The second position is the second power, 4, and so forth. Figure 5-11 gives the first 20 powers of 2.

Note incidentally that this table shows 2^0 as 1. This convention derived originally from positional notation, since the zero digit position always represents *units*, but is also necessary for full expansion of the system of powers and roots.

Now let's take any two of the numbers in the table and multiply them. For instance, 2^5 is 32 and 2^7 is 128. When we multiply 32×128, we get 4096—which is 2^{12}.

But $5 + 7 = 12$. Is it just coincidence that the exponents of the two factors add up to the exponent of the product? Let's try again and multiply 2^8 times 2^6. This is 256×64, or 16,384—which is 2^{14}, and $14 = 8 + 6$.

It's not coincidence. No matter how many times we try it, we will find that it works the same way every time. The reason was discovered back in 1614 by John Napier. He wrote down two series of numbers. One was obtained by counting: 1, 2, 3, 4, 5, 6, 7, 8, and so forth. The other 16, 32, 64, 128, 256, and so forth.

His great discovery was that when the two sequences were put side by side to form a table, the *adding* involved in the counting sequence corresponded to the *multiplication* involved in doubling sequence.

The doubling sequence is the same as the table of powers of 2, because doubling corresponds to multiplying by 2. This means that adding exponents corresponds to multiplication of roots, and offers a way to perform multiplication by doing no more arithmetic than simple addition.

What's more, the inverse operations should also correspond, so that division of roots can be done by subtraction of exponents.

All of this is the case, but that's not the most significant feature provided by the existence of *logarithms*. The use of log tables changes multiplication problems to problems of addition. Since raising to a power is an extension of multiplication just as multiplication is of addition, you might suspect that using log tables would permit raising to a power by multiplying—and extraction of roots by dividing. If you harbor such suspicion, you are correct. And this, in practice, is how we manage to get useful results with powers and roots.

So long as we restrict ourselves to using only the integral powers of a root such as 2—that is, the powers which involve integers as exponents, such as 1, 2, 4, 8, 16, and so forth—the usefulness of the logarithmic approach is somewhat underwhelming. It's fine if you happen to want to multiply 8 times 16, but doesn't help at all if we need to multiply 9 times 15 instead.

Napier's contribution was to extend the idea of an exponent to include rational and real values as well as integers. This extension may well stretch your imagination; since an exponent of 2 indicates that a number appears as a factor 2 times, a rational exponent such as 3/2 would indicate that the number would appear as a factor one-and-a-half times. But how can you picture a number appearing *half* a time?

It's very much like attempting to count a gallon of milk, and stems from the same basic facts. When we try to imagine or picture what happens in mathematics, we do so in terms of the real world. In so doing, we tend to lose sight of the fact which forms a near-constant refrain in these pages—that mathematics does not deal with the real world. These images we form are nothing more than aids to the imagination.

Thus, counting is a physical action in the real world, which maps into and onto the mathematical operation of addition. The picture of counting helped us to understand addition, but we can add things which can never be counted, such as gallons of milk.

Similarly, the exponent is only an operand of the operation called *raising to a power*, which we illustrate and *initially* define in terms of factors and multiplication.

The exponent, however, is not limited to being a member of the set of integers. Nothing in its definition keeps us from using any real number as an exponent, so long as we can make it follow all the rules which apply to exponents as they are initially defined.

Using the easier-to-imagine integer values of exponents for the root or base 2, we know that the power 2 is equal to 2^1, and the power 4 is equal to 2^2. The exponent which would produce 3 as its resulting power could logically be expected to be somewhat between 1 and 2.

By a somewhat involved process which we haven't yet examined, an approximation for the required exponent can be found in 1.5850. We can go through the same process to find what exponent would give us 6 as a power, and find the value 2.5853.

If our correspondence between addition of exponents and multiplication of roots is valid, then $2^{1.5850} \times 2^1$ should equal $2^{2.5853}$ since $3 \times 2 = 6$, and adding exponents would give us $1.5850 + 1.0000 = 2.5850$. We are off from the calculated value of 2.5853 for the exponent giving 6 by only 0.003, and this comes about because we are using only 4-place values for the exponent. In each case, the last digit is where all the error lies, so error in the last digit of the result can be expected.

We call the exponent in this situation the logarithm of the number which corresponds to the power, and the root is known as the base. When we are done and need to go back to the original number from a computed value of logarithm, we need to inverse the function. This is called the *antilog*. Thus 2.5853 is the logarithm of 6 to the base 2, and 6 is the antilog (base 2) of 2.5853.

In practice, we hardly ever use logarithms (abbreviated in math jargon to "logs") to the base 2. Napier originally calculated his table to the natural base which is an irrational real number between 2 and 3 now called epsilon or ϵ; its approximate value is 2.7181. This value is the result of summing the first 8 terms of an infinitely long series of numbers; it took Napier 20 years of effort to develop his tables, for each log had to be produced by calculating similar sums of series.

The natural base is still used in math and physics, but for calculations we usually use logs to base 10 (*common* logs), so that similarities to the positional notation we already know can be retained.

In common logs, the fractional part of the log, or those digits to the right of the decimal (called the *mantissa*), is the same for any sequence of digits no matter where the decimal point happens to be located. That is, the mantissa for the number 1.2345 is exactly the same as that for 1234.5, since both numbers have he same sequence of digits.

Location of the decimal point is given by the figures in the log to the left of the decimal point (called the *exponent*, but this is a different meaning of the word than that used in raising to a power, so be careful). The exponent part of a log to the base 10 tells us what power of 10 to multiply the antilog of the mantissa by in order to get the antilog of the entire logarithm.

In the first example, the exponent would be 0; in the second, it would be 3. Thus, the complete log for 1.2345 would be 0.091403, while that for 1234.5 would be 3.091403.

Tables of logs and antilogs are produced and published for use in doing all sorts of computations. They make it possible to raise a number to the power 2.4653 should you desire, by looking up the log of the number, multiplying that log by 2.4653, then looking up the antilog of the result.

Similarly, you can calculate the 13th root of a number, by dividing its log by 13 and looking up the antilog.

It must be stressed that the result of any calculation done with log tables is not exact, since the logs themselves are approximations. Mathematicians call the deviation in accuracy which results from this source *roundoff error*, since it is introduced by rounding off the logs when they are originally calculated.

In practice, the roundoff error introduced by using logarithms can be safely neglected. Hitting the moon with a spacecraft launched from the earth has been compared to shooting a rifle at a golf ball from a speeding train—but it has been done several times, and all the calculations involved in doing so were made with the aid of log tables. The error, while there, is certainly extremely small.

Functions in General

Now that we have moved from the physical act of counting through the tabulated finite systems of equations which defined the basic arithmetic operations, into the more exotic operators which could not be tabulated, and on to the tabulated logs which resulted, it's time to look for the underlying pattern.

The one thing all of these different operations have in common is that they can, in principle at least, be represented in tabular form.

Even the operation of raising to a power, which cannot be tabulated in a *finite* or limited listing, can be considered an endless table. We made use of part of this table, in fact, when we listed the first 20 powers of 2 in Fig. 5-11.

Similarly, while the operation of counting never comes to an end, every count which is tallied can be entered in a table, so all mathematical operations we have examined to date can be recorded in a tabulation.

Or, as we observed at the end of Chapter 2, almost anything done in mathematics can be described as a *table* lookup process, which in turn is another way of saying *a mapping of sets*.

Another name for the mapping of one set into itself or another set is *function*. The addition tables define a mapping of the set of integers onto itself, so we can with equal accuracy call them the *addition function*. Similarly, every operator we have examined defines a mapping of numbers, and so is also a function.

We have already used the term *function*. Any function which is a one-to-one mapping has an inverse function which reverses the effect of the original. Thus we have addition and its inverse, subtraction; real multiplication and its inverse, real division; and the log function with its inverse we call the antilog.

Some functions which are many-to-one mappings also have partial inverses. We encountered this situation in raising to a power. The inverse function, root extraction, loses some of the original information because the original function is a many-to-one mapping and so cannot be uniquely reversed.

The few operators we have encountered so far, while adequate for most practical arithmetic problems and much basic algebra, are only the barest beginning in the subject of functions.

Higher mathematics abound with functions Trigonometry, for instance, which is the mathematical model behind the sciences of surveying and astronomy, deals with functions which are defined by the ratios of sides of a right triangle.

Beyond trig, we find *hyperbolic* functions, *elliptic* functions, and a host of others.

All, however, are simply defined mappings or operations that help us isolate the patterns which appear in almost all real-world events. We use the mappings which fit the problem at hand; the tedious part of learning math is that of learning enough about all of the functions already defined to be able to pick the one that fits any given problem.

We could continue discussing various functions and look at nothing else in the space remaining in this volume, but that wouldn't help us make any practical use of what we're finding out. Let's turn our attention to another aspect of operations, functions, rules, or whatever you prefer to call the verbs of the mathematical language.

World without End

By now, it should be becoming apparent that the world of mathematics is like a simple spiral, which loops back on itself at every turn yet never returns to the starting point. At every loop, it has taken a step forward as well as around.

The language of mathematicians (as distinct from the language named mathematics) even has a special name to describe this looping-back situation. It's called *iteration*, meaning to repeat the procedure. The everyday word "reiteration" is a redundancy that means "to repeat again."

Iteration

All of the language named mathematics is based on iteration. The lines between math and the real world depend upon the sets of numbers defined by counting, and by the iterative application of the successor function to imitate counting.

Similarly, addition by itself may be enough to solve simple problems, but it won't do for most real-world applications. The result obtained from one operation must be plugged back in as an operand for another in most cases.

Mathematicians use the principle of iteration to achieve rigor when they set out to prove theories. To prove that something holds true for all numbers, they might try to prove it for each number individually. This task could never end, because the sequence of numbers has no limit.

What they do instead is to prove by *induction* using the principle of iteration. They prove the theory for one number n first. Then they prove that if it is true for n, it is also true for $n = 1$ (which they call the *variable of induction*).

When this is accomplished, they have not proved it for just two numbers. They can iterate the proof, so that the second number of the first round becomes the first in the second round; in principle this could go on without limits, until the proof included every one of the paradoxical unlimited set of numbers.

Even the most rigorous of mathematicians, however, does not require that this be done. Once it is proved possible, that satisfies him.

Another application of iteration is the generation of an infinite series. For instance, we saw when we were looking at the principles behind logarithms that they depend upon the correspondence of the series generated by iterating the operation $n = 1$ and that generated by iteration of $2 \times n$. A little later we'll look at some more complicated series.

Factorials

Combining the arithmetic sequence involved in counting and the operation of multiplication in an iterative manner provides us a function called the *factorial* of a number. The factorial of any number is the product of all numbers from 1 up to it. Thus the factorial of 1 is 1, that of 2 is 1×2 or 2, for 3 it is $1 \times 2 \times 3$ or 6, and so forth.

Only integers are involved in the factorial, which is symbolized by an exclamation point; the factorial of 1 is 1!, that of 2 is 2!, and so on. Values of the factorial rise with amazing rapidity. Figure 5-12 lists values for 1! through 10! to illustrate the idea; large factorials are usually estimated rather than calculated, by use of approximation techniques.

The factorial becomes involved when we try to visualize what happens in the shufling of a deck of cards. Although most card decks have 52 cards, 13 in each of four suits, let's use a simpler deck for our first look—you'll soon see why.

If we have four cards in the deck, marked A, B, C, and D, and shuffle them in such a way that each time the four come up in different order, how many different patterns can we have?

N	N!
1	1
2	2
3	6
4	24
5	120
6	720
7	5040
8	40320
9	362880
10	3628800

Fig. 5-12. This table lists the first 10 factorials. Notice how, from a slow start, they grow rapidly. Factorial of 11 would be 11 times 3628800.

Let's see if we can list the possible patterns. The first one which comes to mind is the natural order *ABCD*. Now we can transpose the last two cards to get *ABDC*, and let the *D* card keep moving forward to get *ADBC* and *DABC*. So far, we have four patterns, three of which begin with card A as the first card.

These three are not the only patterns that result when card A comes up first, though. Let's see if we can list the others.

One way to do it is to guess, but this does not guarantee that we won't overlook a pattern or two. The mathematical way to do this sort of thing is to discover some systematic technique which can guarantee that all possible patterns are checked. We can do it by iteration, starting with a deck that has no cards in it. This deck having no cards can produce no pattern.

If we then add one card to the deck, it is the only card in the deck and can only produce one pattern (Fig. 5-13). If it is card *A*, every shuffle of that one-card deck produces the pattern *A*. (Bear with the seeming stupidity of this; we're setting up a starting point and a set of rules for iterating.)

When we add a second card, *B*, to the deck, we can get either of two patterns: *AB* or *BA* (Fig. 5-14). That is, either of the cards can appear at first, but whichever comes first must be followed by the other and no other patterns can be achieved with the two-card deck.

Adding a third card, *C*, gives us more patterns (Fig. 5-15). First we'll put the *C* after the last card of the first pattern of the previous deck to get *ABC*, then after the next-to-the-last card of the first pattern to get *ACB*. Since this brings the new card to position number 2 in the deal, we've exhausted all the

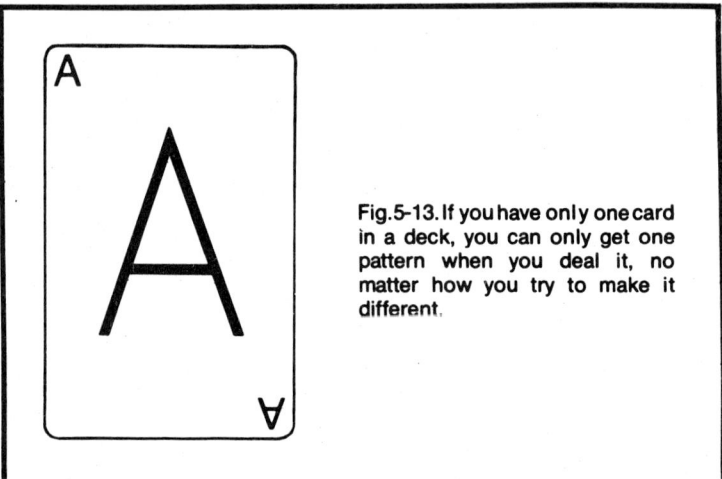

Fig. 5-13. If you have only one card in a deck, you can only get one pattern when you deal it, no matter how you try to make it different.

patterns which can start with *A*, and it's time to rotate. We get *ABC*, *BCA*, and *CAB* when we rotate the first of our new patterns, and *ACB*, *CBA*, and *BAC* when we rotate the second.

We can prove that these six patterns are all that we can get with three cards, because each card appears twice in each position.

Now when we step up to our four-card deck, we have six patterns to insert the D card into. We get *ABCD*, *ABDC*, and *ADBC* from the first. From the second, we get *BCAD*, *BCDA*, and *BDCA*. The third gives us *CABD*, *CADB*, and *CDBA*.

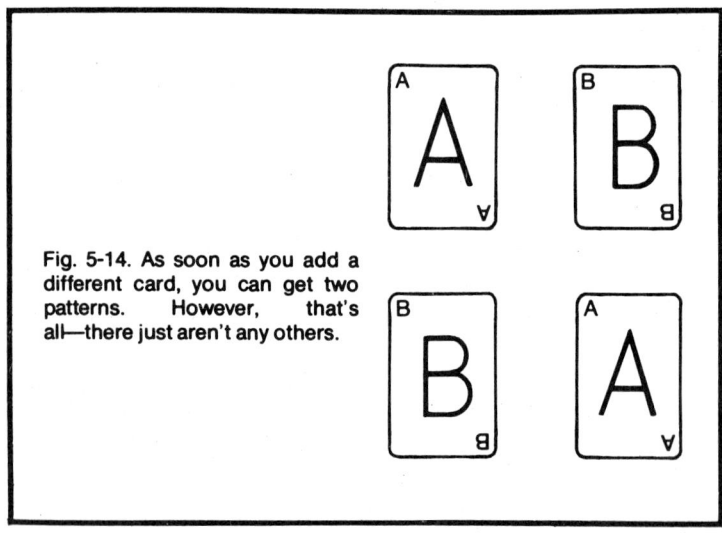

Fig. 5-14. As soon as you add a different card, you can get two patterns. However, that's all—there just aren't any others.

128

From the fourth 3-card pattern we get *ACBD, ACDB,* and *ADCB.* The fifth gives *CBAD, CBDA,* and *CDBA.* Finally, we get from the sixth *BACD, BADC,* and *BDAC.* Now we rotate these.

Rotating the first pattern gives us *ABCD, BCDA, CDAB,* and *DABC.* Rotating the second gives us *ABDC, BDCA, DCAB* and *CABD.* Rotating the third produces *ADBC, DBCA, BCAD,* and *CADB.*

The fourth pattern, *BCAD,* is one of the rotations of the third, so there's no need to rotate it again because we already have all the patterns which could result in our list.

Similarly, the fifth pattern is a rotation of the first and must be omitted, and the sixth is a rotation of the second.

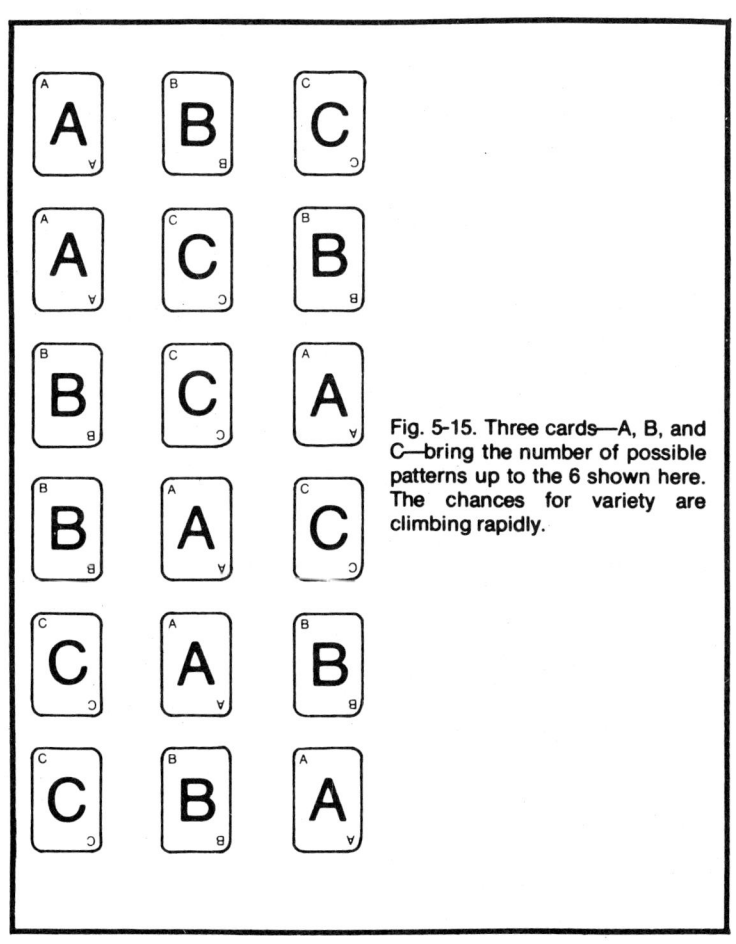

Fig. 5-15. Three cards—A, B, and C—bring the number of possible patterns up to the 6 shown here. The chances for variety are climbing rapidly.

129

Our seventh new pattern, *CABD*, is also a rotation of the second; the eighth and ninth similarly are rotations of the third and first.

Our tenth new pattern, though, has not shown up yet in our listing. Rotating it gives us *ACBD*, *CBDA*, *BDAC*, and *DACB*. Pattern 11 rotates as *ACDB*, *CDBA*, *DBAC*, and *BACD*; pattern 12 produces *ADCB*, *DCBA*, *CBAD*, and *BADC*.

The remaining patterns, like patterns 5 through 9, turn out to be rotations of these, so we end up with 24 patterns (Fig. 5-16) from a 4-card deck. Of these 24 patterns, 6 of them start with card *A*, 6 with card *B*, 6 with *C*, and 6 with *D*.

Notice that when we went to rotate the patterns, we found that all those patterns generated by 3-card patterns which began with card *B* or card *C* turned out to be included in rotations of those patterns which began with card *A*. Thus, to iterate up to a 5-card deck, we need not bother with inserting card *E* into any patterns except those which begin with *A*; this will cut down on the effort involved in our listing.

And how much effort is involved as we do this for larger and larger decks? Let's tally up some totals. For a deck of 1 card, we had 1 pattern. For 2 cards, we had 2 patterns and 1 of them started with card *A*. For 3 cards, we had 6 patterns and 2 of them started with *A*. For 4 cards, we had 24 patterns and 6 of them started with *A*.

How does this compare with the factorial functions listed in Fig. 5-12? Take a look at Fig. 5-17, which compares them.

It looks as if the number of patterns of a deck of n cards is given by the function $N!$, and the number of cycles is given by the function $(n-1)!$.

That means that in a deck of 10 cards, the possible number of patterns would be more than 3 million; the conventional 52-card deck has an astronomical number of possible patterns—more than one person could list in a long session at the card tables.

Numbering Infinity

When we first looked at logarithms, we claimed a correspondence between two sequences—but both of them were infinite series, without limit at their upper ends. How, then, could correspondence be claimed or proven?

The way we did it was to start at the beginning of each of the sequences and show that the first element of one

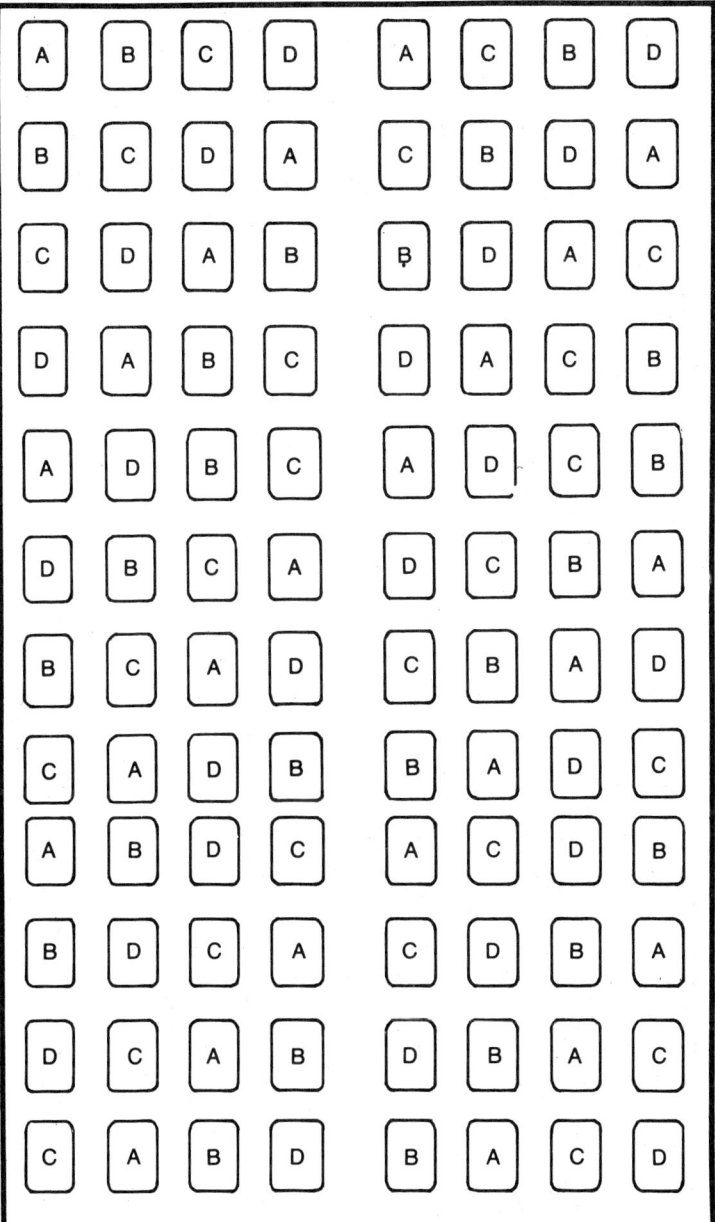

Fig. 5-16. When you have four cards, there are 24 possible patterns on any one deal. Every one of the patterns is different, as you can see, but they fall into one of six **cycles**. Within each cycle, the difference is only a matter of which card comes up first.

NO. OF CARDS	N!	PATTERNS	CYCLES	(N−1)!
1	1	1	1	1
2	2	2	1	1
3	6	6	2	2
4	24	24	6	6

Fig. 5-17. This table compares the kinds of cards, factorials, number of patterns, and number of cycles. A system emerges.

corresponded to the first element of the other, the second of one to the second of the other, the third to the third, fourth to fourth, and so on.

By induction, the correspondence is carried on as far into the series as you like.

Similar tactics are used whenever it becomes necessary to deal mathematically with *infinity*, which is not a number, but a place beyond all possibility of definition. The simplest way of reading the infinity symbol ∞ in math is as *undefined*, but occasionally it's necessary to take it into account.

The places where it becomes necessary to meet the idea of the infinite face to face and master it, are in *progressions* and *series*. We've already met the arithmetic progression of counting, which generates the series 1, 2, 3, 4, 5 ... (the 3 dots are called an *ellipse*—they indicate that the series goes on without limit, or is infinite). And we've seen the geometric progression of doubling which generates the different infinite series 1, 2, 4, 8, 16, 32 ...; these are only two of an unlimited number of possible sequences.

Should you want to, you could write an operator symbol between each term in an infinite series and perform the indicated operation over any limited part of the sequence. For instance, you could use the addition operator on the geometric series to form another series: "1 + 2 + 4 + 8 + 16 + 32 + ..."

If you take the sum of the first 2 terms of this series, you get 3. The first 3 terms sum up to 7. The first 4 total out to 15. Each of these totals is 1 less than the next term in the series; 3 is 4−1, 7 is 8−1, and 15 is 16−1.

We can write the sum of the first $n - 1$ terms of this series then as $2^n - 1$. Doing so permits us to examine what happens to the sum as we keep increasing the value of n. That is, we can write this sum as an equation: $y = 2^n - 1$, and study what happens to the dependent variable y as we vary the independent variable n over the range from 1 up as far as we like.

When we do so, we will find that the value of y increases without limit, since it depends upon the value of the nth term in the geometric progression, which increases without limit as the progression proceeds.

Series which behave this way are said to be *divergent*.

Let's try now a series which is not divergent, generated by taking the *reciprocal* of factorials. A reciprocal of a number is equal to 1 divided by that number, so that as the number itself gets larger the reciprocal gets smaller. The specific series is the sum of the reciprocals of the factorials of the counting sequence:

$\epsilon = 1/0! + 1/1! + 1/2! + 1/3! + 1/4! ... + 1/n! + ...$

The first two terms of this series are both 1; factorial zero is defined as 1 in order to make much of higher math consistent, and $1/1!$ is $1/1 \times 1$ or 1 by ordinary arithmetic.

Since the nth term of the series is equal to the reciprocal of $n!$, which gets larger without limit as n increases, this series can never reach a total of 3 no matter how many terms we include. In fact, since each term is much smaller than the term preceding it (after the third or fourth is reached), the error involved in cutting it short cannot be more than the value of the last term we include in our calculations.

When we total the first 8 terms of this series, we get the results which are tabulated step by step in Fig. 5-18. This value is the approximation of the base of natural logarithms that we met earlier in the chapter, which is known today as *epsilon* or ϵ; it can be calculated as accurately as you like by using more and more terms of this series.

While it might seem that the value must always increase as more terms are taken (a true observation), the fact remains that the increase is getting smaller each time. Thus, some upper limit exists, which the number cannot exceed no matter how many terms are used from the series.

This limit is the real number represented by ϵ; all values for it that were calculated by summing terms of the series are rational approximations to it, which are less than the real number itself.

TERM	PREVIOUS SERIES SUM	VALUE OF THIS TERM	SUM OF SERIES TO DATE
1	0	1/0! = 1	1
2	1	1/1! = 1	1 + 1 = 2
3	2	1/2! = 0.5	2 + 0.5 = 2.5
4	2.5	1/3! = 0.166667	2.5 + 0.166667 = 2.666667
5	2.666667	1/4! = 0.041667	2.70833333
6	2.70833333	1/5! = 0.008333	2.71666667
7	2.71666667	1/6! = 0.001388	2.7180554
8	2.7180554	1/7! = 0.000198	2.7182538

Fig. 5-18. How to sum up an infinite series. This is a calculation of the value of **epsilon**, an irrational real number defined by a series, using the first 8 terms of the series.

The series is thus said to converge to this limit, or to be *convergent*; series which are divergent have no limit to reach.

Most of the functions which we mentioned earlier in the chapter are calculated by summing such series. The advent of the computer made the calculation much simpler, and we now have accurate mathematical tables of which the giants of the 19th century could only dream.

The study of series calculations is a subject in itself, forming the gateway to *analysis*, one of the more important branches of applied mathematics. At this point, we must turn to applying what we have discovered rather than continue to sum series.

From Operation to Operand

The idea of iteration, as we have observed, is basic to all mathematics. The result of an operation is, itself, an operand which can form part of another operation.

This fact permits us to use more than one operation in a problem, and we've been making use of it unobtrusively all the way through this book. For instance, when we sum the series 1, 2, 4, 8, 16 to get a total of 31, we first do the operation $1 + 2 = 3$, then use the result of this in the operation $3 + 4 = 7$, plug this result into $7 + 8 = 15$, and finally do $15 + 16 = 31$. A total of four operations were involved to sum up 5 terms.

When we're summing up a series it makes no difference which order we use for the summation. We could have done it as $16 + 8 = 24$, $24 + 4 = 28$, $28 + 2 = 30$, $30 + 1 = 31$ and the answer would have been the same. Sometimes, however, the order in which the operations are performed does make a difference.

When it does, we indicate the required sequence by using parentheses, brackets, and braces. The first sequence of summations would be written like this: $\{[((1 + 2) + 4] + 8\} + 16$. This tells us to first add $1 + 2$, then use the result as the operand for adding to 4, and so on.

Similarly, the second sequence of additions would be written $1 + \{2 + [4 + (8 + 16)]\}$, which indicates adding $8 + 16$ first, then adding the result to 4, and so forth.

When we're adding it makes no difference. When we mix operations, such as addition and multiplication, it does. Consider $1 + 2 \times 3$. If we take the operators in left-to-right sequence, we get $1 + 2 = 3$, followed by $3 \times 3 = 9$, and have 9 for our result. Doing the operations right to left gives us $3 \times 2 = 6$, followed by $6 + 1 = 7$, for a result of 7. Which is right?

To prevent such confusion, mathematicians established a "rule" that in arithmetic, when no parentheses are present multiplication and division must be done before (take precedence over) the addition and subtraction. Similarly, raising to powers takes precedence over multiplication or division. When parentheses are present, operations inside the parentheses take precedence over those which are not.

If the sequence of operations established by this rule fails to fit the problem at hand, the desired sequence is established by using parentheses. Thus without parentheses, $1 + 2 \times 3$ would always give $2 \times 3 = 6$, $+ 1 = 7$. If the problem really required that the answer be 9, it would be written as $(1 + 2) \times 3$.

Note that the rule is arbitrary; it could have been done any other way and would have worked just as well. Now, however, everyone who uses it follows it, so you can't go the other way and use the same equations everybody else does. The choice has already been made, and now we can only follow it.

Chapter 6
Putting Math to Work

In the preceding five chapters, we saw that the language called mathematics is used to make statements about operands, and that operations may be used to map two operands into a single new operand.

We also saw that some of these operations are themselves mappings of the real-world action called counting. These mappings make it possible for the results of the operations to also be related to real-world situations, and that's what makes math useful.

In order to make practical use of mathematics, we have to now learn how to map real-world situations into a *mathematical model* or, to use more conventional words, we must learn how to state our problem in the language of math.

One of the simplest such mappings is the sort we met early in our educational careers: If you have one apple, and I have another, and if both of us place our apples on an empty platter, how many apples will be on the platter?

The operation required for this mapping is the one known as addition; in words, we could state the problem as my apple plus your apple equals how many?, but the teacher writes it on the board as $1 + 1 = 2$.

Few real-life problems map into their mathematical models quite so simply and directly. In this chapter, let's examine some possible real-world problems and see how we can construct mathematical models for them.

And remember while we do this that once the proper model is established, the problem is solved. Getting the answer from the model is a purely mechanical process, even if it turns out to be tedious at times. That's why computers can work; the answers are built into the problems given them.

DESCRIBING A PROBLEM

The heart of mathematics' applicability to real problems consists of the correspondence established between the operands of mathematics and the objects in the real problem. This in turn means that the whole secret of building mathematical models consists of establishing the proper correspondences.

But it's not just the correspondence of object and operand that we have to be careful about. We must also be sure that we pick the proper *operations* for use in our model, so that they correspond correctly to the *actions* involved in the real world.

For instance, had we chosen for our operation *subtraction* rather than addition, we would have come up with the mathematical model $1 - 1 = 0$, which would not have corresponded to the real-world count of two apples on the platter.

Most of the mathematical training we get in conventional approaches to the subject involves practice in using the various operations. Very little of the training prepares us to *choose* the right operation before using it.

That little bit which does intend to show how to choose the operation—the so called *thought problem*—is often skipped over by technique-oriented teachers, or eliminated by the pressures of an overcrowded schedule.

Let's start our study of mathematical modeling, then, with a real-world situation most of us meet every day. First we'll see how it is actually handled in real life, then we'll look at some alternate ways of doing the same thing mathematically to find out how it could be done.

Making Change

Back in Chapter 5 we touched upon the change-making problem while examining the operation of division. Now let's look at it in some more detail.

Let's assume that we have bought something which costs 21 cents, and have handed the sales clerk a dollar bill.

What will probably happen is that the clerk will place the dollar bill on the cash register ledge below the keyboard, count out change from the drawer, place the bill in the drawer and close it, then repeat the change count while giving us our change.

That count will be *twenty-one cents* to tally the item itself; *twenty-two, twenty-three, twenty-four, twenty-five,* as four pennies are counted out to bring the total to the value of a larger coin; *fifty,* as a quarter joins the four pennies and brings the total to the next larger coin; *...and fifty makes a dollar,* as a half-dollar completes the process.

If you are in an area where half-dollars are not in fashion, the count may be *fifty, seventy-five, one dollar* as three quarters rather than a half-dollar and quarter join the four pennies.

Perhaps your store uses one of the new change-calculating cash registers. In this case, the clerk will ring up the 21-cent sale, then ring up the one-dollar offered, and the machine will indicate that the change due is 79 cents.

The clerk may then pick out three quarters and four pennies, but the odds are that your change will still be counted out to you just as if the machine hadn't calculated it.

The counting technique is a tradition honored by time, and based on the solid conviction that counting is more reliable than most clerks' mathematics. Still, this gives us a chance to figure out several mathematical models which can be used to calculate the proper change.

The first step in building a mathematical model of the change problem is to determine how much change is required. The sale was for 21 cents, and the payment was one dollar. How much change should there be?

Pennies and dollars both refer to money, but they're different units, and our mathematical operations require that each operand refer to the same kind of unit. Thus, even before we can take the first step of our model we have to map our real-world money into mathematical units.

The clerk does this without thinking, but a computing machine would have to be told that one dollar is equal to 100 cents. The first statement in our model, then, will be the equation $\$1 = 100¢$.

Now we can map the change calculation. The question is, how much more than 21 cents is a dollar? This maps into the equation: *change* = 100 − 21, which becomes our second statement.

We perform the subtraction operation at this point if we are familiar with arithmetic. Since our intention is not so much to make the change as it is to explore in detail the building of a mathematical model, we'll do it differently; our third statement will be a grouping of both *figures* into *tens* and *units* to read: *change* = (10 − 2) *tens and* (0 − 1) *units*.

We will mechanize the *borrow* required in multidigit subtraction in the fourth statement: *change* = (9 − 2) *tens and* (10 − 1) *units*.

Those operations can be taken to the decimal subtraction table (Fig. 5-5) to yield the fifth statement: *change* = 7 *tens and* 9 *units*, which can then be grouped back into positional notation to provide statement six: *change* = 79 *cents*.

Now we're at the same point that the calculating cash register reached. The difference is that we know exactly how we got here: it's a chain of six equations, which included two table lookups.

We do not yet have the change, though. At this point we only know how much is required.

We could provide the change in any of a large number of ways. We could, for instance, count out 79 pennies. Another way would be to give four pennies and 15 nickels. Still another would be to give four pennies, one nickel, and seven dimes.

One nickel and 74 pennies would do it, as would one dime and 69 pennies. Let's try to do it with the smallest possible number of coins, though. This is the way most clerks do in real life, unless they run out of some size of coin and are forced to make do with what's on hand.

One way of using the smallest possible number of coins is to count down from the total change using the largest possible size coin at each step in the counting.

This would mean that first, we would us a half-dollar. Subtracting this from the 79-cent change figure would leave 29 cents' change to go. Next, we would use a quarter, since that would then be the largest possible coin for this step. Subtracting it would leave four cents. The largest possible coins for this would be pennies, and it would take four of them to finish the job.

This is essentially what the clerk does with the counting, without having to calculate the total change first. It is also essentially the technique of integer division shown in Fig. 5-7, except that the divisor changes whenever the remainder is less than the divisor.

Another way of comparing this action to integer division is to say that it is repeated integer division, using first the largest coin size (half-dollar) as the divisor, and working through each size in turn until reaching pennies.

That is, we first divide 79 by 50 (number of cents in a half-dollar) to get a quotient of 1 and remainder of 29.

We then divide 29 by 25 (cents per quarter) to get quotient 1 and remainder of 4.

Now we divide 4 by 10 (cents per dime) to get quotient 0 and remainder 4; also by 5 (cents per nickel) with the same result. Finally we divide 4 by 1 (cents per penny) to get quotient 4 and remainder 0.

To pick up the change, we take 1 half-dollar, 1 quarter, 0 dimes, 0 nickels, and 4 pennies.

To sum up the mathematical model of the change problem, here are the equations involved:

1. $1   = 100¢
2. change   = 100 − 21
3. change   = (10 − 2) tens and (0 − 1) units
4. change   = (9 − 2) tens and (10 − 1) units
5. change   = 7 tens and 9 units
6. change   = 79 cents
7. halves   = 79/50
8. quarters = (rem 79/50)/25
9. dimes   = (rem step 8)/10
10. nickels = (rem step 9)/5
11. pennies = (rem step 10)/1

By inserting several more division steps between equation 6 and 7, to handle the situation when the amount of change is larger than a dollar, this model can be made to fit all change-making situations.

That, of course, is the real usefulness of any mathematical model. Once it is built, it can be applied to all similar situations rather than being applicable to only one case. In this case, the model turns out to be more trouble than it's worth for humans—but automatic change-making machines have been built in which a similar model was built into hardware to eliminate the need for a human to count out change.

Checkbook Balancing

Another money-based situation which most of us encounter in real life (and do without worrying about construction of a suitable mathematical model) involves keeping track of the money in our checking accounts.

The situation is simple enough. You deposit money in your checking account in a bank, and then write checks against the deposit in order to keep from having to handle cash in large amounts.

The problem comes when we write checks for more money than the account contains at any instant. Banks, for some reason, take a dim view of such occurrences. So do the recipients of the checks, when the banks send them back stamped *insufficient funds*.

To prevent such embarrassment, it's necessary to keep the account balanced. You do this by adding to the balance every deposit, and subtracting every check written.

The actual real-life actions correspond to counting up and down, using money instead of counts. Every deposit to the account corresponds to a count up of that many pennies; each check or service charge is a count down.

We know that the mathematical operator of addition maps into the action of counting up, and that of subtraction into the action of counting down, so we add deposits and subtract checks.

It works like this: Suppose you start by depositing $100 to open the account. Then you write one check for telephone service in the amount of $9.68 and another for groceries in the amount of $35.19.

At this point you deposit another $150 in the account. Your automobile breaks down and must be fixed, which forces you to write a check to the garage for $128.50. Now how much do you have left in the account?

Checkbooks have, to help you keep track of things, check logs built into them. Figure 6-1 shows a typical log for the account we have been discussing.

When the account was first opened with a $100 deposit, there was no prior balance, so the first line of the log shows the deposit. The balance is the old balance (0) plus the deposit ($100), for a total of $100.00.

When you wrote check 1 to the telephone company, the amount was $9.68. Since this was a check rather than a deposit,

NO.	DATE		AMOUNT		BALANCE	
	Jan. 23	Deposit	100	00	100	00
1	Jan. 25	Telephone Co.	9	68	90	32
2	Jan. 26	Grocery	35	19	55	13
	Jan. 30	Deposit	150	—	205	13
3	Jan. 30	Garage	128	50	76	63

Fig. 6-1. Account book furnished as part of checkbook with many bank accounts would look like this after being filled out to record checks as described in text.

the new balance is equal to the old balance ($100) minus the draft amount ($9.68), for a balance of $90.32.

Check 2 to the grocer, is subtracted the same way, leaving a balance of $55.13.

Now comes the second deposit. This is added rather than subtracted, since it is a deposit rather than a check, and the resulting balance is $205.13.

The auto repair takes away $128.50 of this, leaving $76.63 in the account.

You can see that this countup/countdown action is assurance that the balance always reflects the uncommitted amount left in your account. So far, so good.

In the real world, however, the checks do not always arrive at the bank in the order you wrote them. For instance, the phone company may have deposited the check you wrote to them in another bank in a distant town, so that it is delayed by several days in getting to your bank.

As a result, when you receive the bank's version of your account, the *statement*, it may look like Fig. 6-2 rather than Fig. 6-1. According to this statement, you have $85.81 in the bank as of the end of the month despite a 50-cent service charge which isn't included in your checkbook figures yet. And, your own checkbook balance shows you have only $76.63.

The difference, of course, is that your checkbook has already had the telephone company's check subtracted when it was written, while the bank statement doesn't show it because it has not yet been paid.

And this is where the mathematical modeling comes into the situation. You have to keep in mind that the statement the bank sends you is a model of their actions regarding your account, while the figures in your checkbook reflect your own actions with the account—and these are two distinct sets of actions. The bank cannot pay an item out until sometime after you wrote it, and the chances are fairly good that every statement will have several such items caught in the middle.

To verify the accuracy of both your own figures, and those of the bank, it's necessary to *reconcile* both sets of figures. To do this, first subtract the service charge shown on the statement from your checkbook balance, so that all the *paid out* entries on the bank statement have corresponding entries in your checkbook. This brings your balance down to $76.13.

Now subtract, from the bank's balance of $85.81, the total of all checks you have written which have not yet been paid. To find out which ones these are, you have to go through the statement and check off each paid-out amount. Matching the figures shows you that the January 27 paid-out item must have been check 2, and that for Jan. 31 was check 3, so only check 1 is still outstanding against the account.

When we subtract the $9.68 amount of check 1 from the bank's balance of $85.81, we get $76.13, which is 50¢ over the balance we have listed in the check log. We see that the bank has taken 50¢ from our account as a service charge so we subtract that from the balance shown in the check log.

Since this now matches the balance shown in the bank statement, we have reduced the two mathematical models to a tautology ($A = A$) and thus they are satisfactorily reconciled.

DATE	TRANSACTION		BALANCE
		Previous	0.00
Jan 23	Deposit 100.00		100.00
Jan 27	Paid out 35.19		64.81
Jan 30	Deposit 150.00		214.81
Jan 31	Paid out 128.50		86.31
Jan 31	Service charge 0.50		85.81

Fig. 6-2. Statement furnished by bank at end of month could look like this for same account and transactions shown in Fig. 6-1. Both are correct; the differences reflect time differences between writing of checks and presentation at the bank.

From the mathematical-model point of view, what you have done is to take the model which reflects your own actions, modify it to include the bank's actions by including the service charge. and compare it to a model of the bank's actions which has itself been modified to include anticipated actions for checks which are out but not yet paid.

So far, the only difference between this mathematical structure and the change problem of the previous subsection is the real-world situation. Both of them are simple situations involving only counting up and counting down.

The checkbook balancing situation, however, is not limited to the simple countup/countdown action. Suppose, with the $76.13 balance, you find it completely necessary to write a check for $90.00.

This means that you would be spending more than you had in the account. Banks do not appreciate such actions on the part of their depositors; in many states, in fact, it's a criminal act to do such a thing. However, there's nothing in the mathematical model to keep you from doing it anyway. Now when you subtract the $90 check from the $76.13 balance you come up with a negative balance. Your account is $13.87 in the hole, or overdrawn.

If you can make a deposit of this much or more before your overdraft reaches the bank, the account itself need never show a negative balance. In this case, the bank won't know you violated its rules—but it's seldom safe to count on being able to make such a deposit in these days of computerized forwarding of checks from bank to bank.

Your own checkbook would show a negative balance that you'd write as −$13.87 after deducting the $90 check. Now, assuming you do make a $20 deposit in time to cover your overdraft, the new balance becomes $6.19; the rest of the deposit went to cancel the negative balance and bring it back up to zero.

This is the closest situation you can reach, in real life, to the mathematical model mapped by the negative numbers. We are still dealing in the set of integers, counting up and counting down, but the bank balancing can go through zero to negative balances. When you're making change, you stop at zero.

The entire monetary system, not only of this nation but of the world, is based on the close correspondence between the mathematical actions of addition and subtraction, and the real-life actions of counting up and counting down.

It works so well that a number of systems intended to make actual use of money a thing of the past are now in field trials. If such systems gain widespread acceptance, you'll have to know how to relate the math to the reality in order to find out whether you have any money.

Calculating Gas Mileage

Both the situations we've examined so far in this chapter have dealt with money, using the set of integers. The first used only the set of natural integers, while the second used all the integers, both positive and negative. Let's look now at a case where the model requires use of the rational numbers.

The worldwide energy shortage has renewed interest in getting the maximum number of miles of driving for each gallon of gasoline used. Interested drivers have always wanted to know what kind of mileage they were getting, since it's one of the key measures of efficiency for both car and driver.

Calculating gas mileage involves the construction of a surprisingly complicated mathematical model. You do not, for instance, just put a measured gallon of gas into the car and see how far you can go before running dry.

You could, of course, and the annual gas-mileage contest sponsored by one of the bigger oil companies does it just that way, but that tells you only how far you could go on *that* gallon. To be realistic, what you really want to know is *What mileage am I getting day in and day out?* or What is my *average* mileage?

Some of the factors affecting gasoline mileage include air temperature, gasoline temperature, humidity, road surface conditions, and driver habits. When you build a mathematical model from which to compute average gasoline mileage, you have to come up with one that takes all these factors into account.

The simplest model which includes all these variables is the standard technique for calculating "average" or "arithmetic mean" of anything. First accumulate the total, and then divide.

If you're figuring the average price of sugar over a 10-day period, you add the price per pound each day (say 59¢, 59¢, 60¢, 50¢, 58¢, 58¢, 60¢, 60¢, and 61¢) to get the 10-day total ($5.85 in this example), then divide by the number of days to find the average price per day (58.5¢).

Similarly, to find the average number of miles per gallon your car gives you, add up the total number of miles driven (say 262) and divide by the number of gallons required to go that far (say 10 gallons) to get the mileage: 26.2 miles per gallon in this case.

However, if you calculate only over single tankfuls you will find that the mileage varies widely from one tank to the next. Part of the problem is the list of variables we already met: in wet weather, you'll tend to get less mileage because of the wet roads, but the increased humidity at the same time may help some.

Another part of the problem is that you don't really know exactly how much gas it took to drive the distance. The normal way to make a mileage check is to fill the tank brim full, record mileage from the speedometer, drive until the tank is nearly empty, then fill back up to the brim to determine how much gas it took. The distance is obtained by subtracting the speedometer reading at the first fill-up from that at the second.

Many folk, though, round off gas-pump figures to the nearest tenth of a gallon. Mileage figures are often read only to the nearest mile. These roundoff errors can introduce large errors into your calculations so long as the total number of both gallons and miles is relatively small.

If you accumulate the totals until they reach fairly large figures, though, the errors in each tankful will tend to cancel. If one reading is a little lower than the true amount, the next may be a little larger, so the errors correct each other.

The mileage reading is automatically accumulated by your speedometer; so even if you do round off to the nearest mile, the error introduced by doing this is much less for a 1000-mile total than it would be for a 10-mile total (0.1% compared to 10%).

By accumulating the totals to large figures, the effects of error are automatically reduced. At the same time, the variables introduced by weather, road conditions, and driver mood also tend to average out so that the figure you come up with is truly representative of your car and your driving.

What's more, once you start accumulating totals to give you a long-term average for mileage, you can then figure the short-term average over each tankful of gas, and see how it differs from the long-term average which has less error in it.

This will alert you to any sudden changes of auto condition, early enough to possibly save gasoline (and maybe even reduce the repair bill by minimizing any damage). That's in the real world, though, and so outside the scope of a discussion on mathematics—just one of the ways math can help with real-world problems.

MORE COMPLICATED DESCRIPTIONS

In the three examples of mathematical-model building we've examined so far, the models have been relatively simple. The first, making change, was a nearly direct mapping from simple counting (and, in fact, is normally handled in real life by counting rather than by mathematics). The second, balancing a checkbook, only extended the model to include negative numbers. The third, calculating gasoline mileage, included use of rational numbers by introducing division.

Almost all applications of mathematics require more complicated models than we have yet built. Usually, you have to find some type of pattern in the real-world problem to be solved, then select a similar pattern in the ideal world of mathematics to determine what kind of model to construct.

Since all mathematical patterns are simplified in comparison with real life, this pattern-matching can become difficult. What makes it such a problem is usually the fact that any of several mathematical patterns can be applied equally well to the real-life situation, and the difference in the results achieved may or may not be significant.

Only experience can show you which pattern should be chosen in such cases—and even experience is not infallible. Some of the most exciting scientific discoveries in our history have been the result of choosing an unconventional pattern, and then following it through to the inevitable outcome. We can expect such surprises not only in science, but in any other application of the mathematical method to real-world situations—which is not much help to you, as a student, trying to pick one of several potentially useful patterns to apply to a single problem.

Let's look at some problems which require more complicated descriptions than those we have already built models for, and see if we can determine which math patterns to use for them. Even more important, we'll try to determine *why* the patterns we pick are the ones that should be picked for these problems.

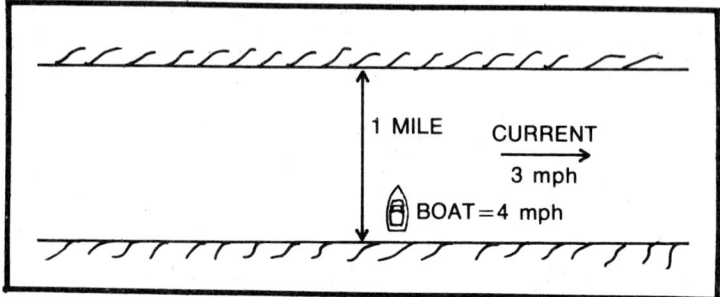

Fig. 6-3. How to get across river in boat, despite current, forms a situation for a problem. To keep the problem simple, we'll ignore such real-life things as the time it takes to get up to speed, and variations of river current along banks.

Crossing a River

For starters, let's take the problem of crossing a river in a motorboat. The current in the river is 3 miles per hour, the river is one mile wide, and our motorboat is acting up somewhat and will only go 4 miles per hour.

From this set of facts, we can draw several problems connected with the crossing. For instance, if we steer straight toward the other side as we cross, the current will carry us downstream while we are crossing and we will reach the other bank some distance downstream from our starting point. One problem, then, is to find out how far downstream we will be carried by the current.

If we want to go straight across despite the current, we can do so by steering upstream rather than straight, so that the current carries us back just enough to balance things out. Doing it this way requires that we steer toward some point upstream; it also means that the crossing will take longer. That gives us two more problems: At what point must we aim to end up straight across? and How long will it take to make such a crossing?

Let's tackle that pair of problems, because it's going to require that we build not one but several models, and solve several problems in turn to come up with a final pair of answers.

The basic situation involved in the problem is shown in Fig. 6-3. Current is flowing from left to right, and we are on the side of the river nearer the bottom of the page in the illustration.

If there were no current at all, the time required to cross the river would be determined entirely by the speed (4 mph) and distance (1 mile), so we could get across in ¼ hour, as shown in Fig. 6-4. The model for this calculation is derived directly from the expression of speed as *miles per hour*; the inverse of this would be *hours per mile*, so the time necessary to go a mile would be ¼ hour per mile. We multiply this time by the distance, 1 mile to get the ¼-hour result.

This, however, ignores the effect of current. When we take the current into effect, we find that it carries us downstream but does not slow us down at all, so it still takes the same ¼ hour to get across. However, for every four feet we go forward at 4 mph, we are carried downstream three feet by the 3 mph current. This is the ratio of the two speeds, 3 mph to 4 mph. This model is set up by observing that the *time* spent in the river is the same for both forces, so the distances would depend only upon the speed.

Since the river is one mile wide, the current would carry us ¾ of this or ¾ mile downstream in the 15 minutes it takes to get across.

An alternate model for determining the distance the current takes us is to use the 15-minute time and the 3 mph speed, then multiply 3 miles per hour by the ¼-hour time. We still get ¾ mile as the distance. Either model fits.

Figure 6-6 shows a model of the two forces acting on our boat at the same time. The circle represents the 4 mph force that can be exerted by our motor. It remains the same no matter what direction we try to go. The 3 mph arm of the triangle represents the force of the river current.

When we steer straight across, as in Fig. 6-5, the force of the current acts at a right angle to that of the motor. The

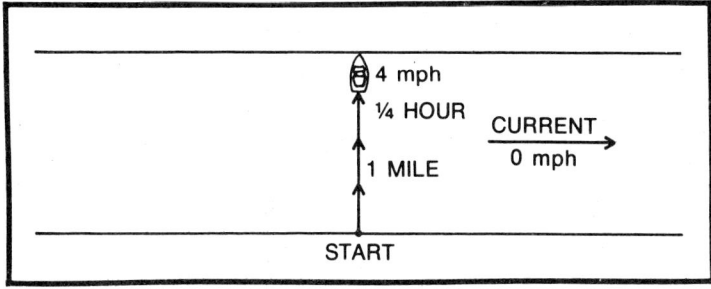

Fig. 6-4. If the current is not present, it would take ¼ hour to get across the mile-wide stream at a constant speed of 4 miles per hour.

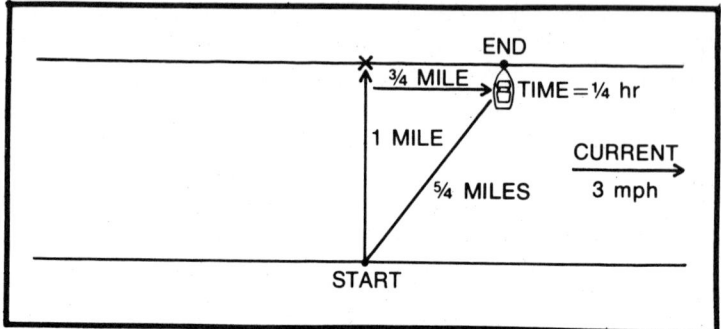

Fig. 6-5. With the 3 mph current, it will take no longer if we don't try to go upstream at all. However, the current will carry us ¾ mile downstream in that ¼ hour it takes us to get across—and that's not what we want to have happen.

effective force acting on our boat is represented by S in Fig. 6-6, and as you can see is greater than either force alone.

The line S also happens to be the hypotenuse of a right triangle, which has sides proportional to forces of 4 mph and 3 mph respectively. Pythagoras' theorem tells us how to find the

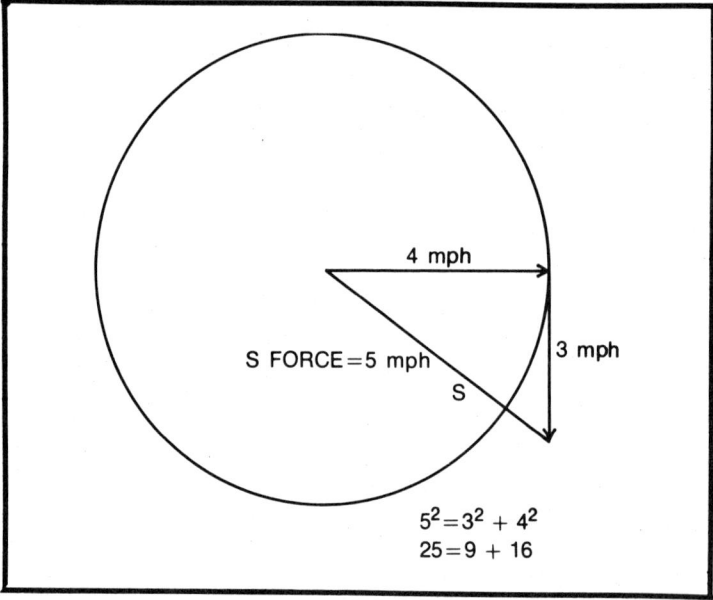

Fig. 6-6. Here's the mathematical version of being carried downstream. The circle represents our boat's force, the vertical line is the force of the current, and the diagonal shows the result when both are applied to the boat at the same time.

length of the hypotenuse of such a triangle if we know the lengths of the sides: If A and B are side lengths and C is that of the hypotenuse, the $A^2 + B^2 = C^2$, or C is equal to the square root of $(A^2 + B^2)$.

This is worked out step by step in Fig. 6-6, and we discover that the effective force S is 5 mph. Of this, 4 mph came from the motor and the other 1 mph came from the force exerted by the current.

The problem we set out to solve, however, was to determine how to steer the boat in order to end up directly across the river, and to find out how long the crossing would take. All we have done so far is to discover some models which can be used to determine the effective forces acting on the boat.

It would be tempting to use the ¾ mile offset we discovered (Fig. 6-5) and decide that we could aim for a point ¾ mile upstream of our target. We might reason that for every 5 feet we go in the direction the boat points, the current pushes us sideways 3 feet, giving us a total effective movement of 4 feet toward the real target. This model is shown in Fig. 6-7.

Unfortunately, it's not a correct model. The falsity of it can be seen by assuming that the boat's speed were to be reduced to 3 mph, the same as the river current. If this happened, the model shown in Fig. 6-5 would change only to make the offset one mile rather than ¾ mile, and that in Fig. 6-6 would give us a speed of about 4.24 mph effective. However, when we aim upstream, we cannot keep from being carried downstream by the current. If we point straight upstream, the

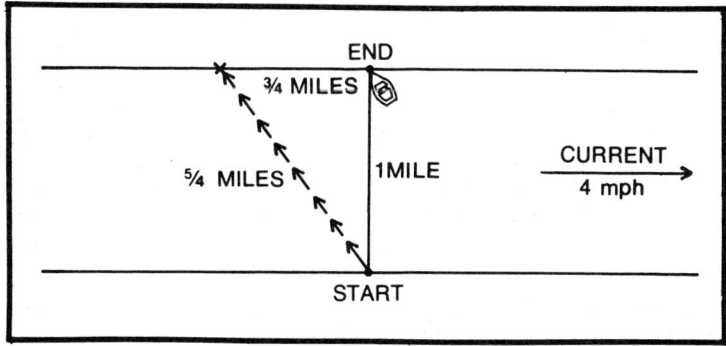

Fig. 6-7. This is an attractive way to approach the problem, but it's wrong. Simply aiming ¾ mile upstream to cancel out the ¾-mile drift downstream won't get the job done.

151

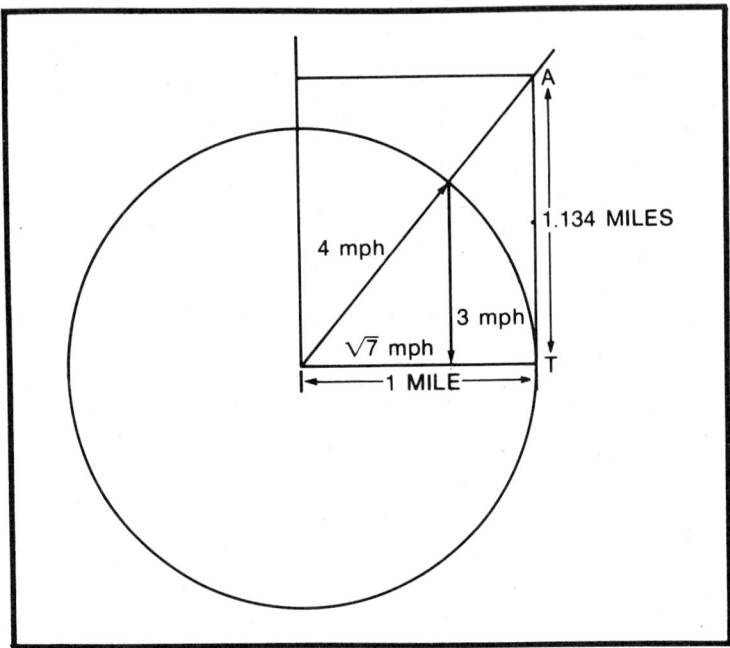

Fig. 6-8. Another circle diagram shows what we have to do. The diagonal, now, is our boat's motor, and the current is still in the same direction. Since we're bucking the current this time, it slows us down, and our effective speed across the river in miles per hour is only **square root of 7**. Since it's a mile across, we have to aim 1.134 miles upstream, and it's going to take us longer to get there too.

forces of the current and the motor will cancel each other and the boat will stand still. If we point to the side at all, the force of the current will be greater than that of the motor and we will be carried downstream. When the forces of the current and the motor are equal, there just isn't any way to keep from being carried downstream and at the same time make any progress toward getting across.

When the motor's force exceeds that of the current, though, we can do it. To see how the applicable model works, let's redraw the force diagram of Fig. 6-6 to take into account the changed requirements. Figure 6-8 is the modified version.

The major difference is that the 4 mph force of our motor is now represented by the hypotenuse of the triangle rather than being a side. The 3 mph force of the current is still a side, but our unknown effective force in the direction we want to go is the other side.

When we apply Pythagoras' theorem again, we find that the effective speed across the river is equal to the square root of 7 miles per hour. The square root of 7 happens to be a real number, not a rational one, so we can only approximate it in our calculations. We'll use 2.65 as our approximation (this number is really the square root of 7.0225, but that's close enough for our purposes).

If it takes us ¼ hour to cross the river at 4 mph, and ½ hour to cross at 2 mph, then it should take 1/2.65 hour to cross at 2.65 mph. The time to cross, then would be approximately 22.6 minutes.

Our aiming point, initially, would be the point shown in Fig. 6-8 as A. We can calculate the distance upstream by setting up ratios of distance/force: $1/2.65 = A/3$. This equation can then be multiplied by the equation $3 = 3$ to give $3/2.65 = A$, and the division then performed to learn that point A is about 1.132 miles upstream.

In actual navigation, the angle between the target point T and the offset aiming point A would be determined by trigonometry (it's a little greater than 48 degrees, 35 minutes—the angle whose sine function is equal to ¾ or 0.75000), and the boat would be kept at this angle to the target point while crossing.

However, in actual real-life experience, the situation would not be so simple as in our example. For instance, we assumed that the current was exactly the same all the way across the river, that the boat had no inertia and could exert its full motor force instantly upon starting, and that we as passengers could withstand such extreme acceleration (going from 0 to 4 mph in zero time would be infinite acceleration, smashing us flat against the stern). None of these assumptions is really true.

Had we attempted to include all these effects, though, our problem would have been far too complicated to use as an example at this point. The false assumptions mean that our answer is not perfectly correct—but they do not invalidate the methods by which we reached our results.

Those methods were: first, to find some known facts of the mathematical world which could be made to correspond to the real-life problem (in this case, Pythagoras' theorem), and second, calculate results from the correspondence.

Finally, we checked by trying other cases, to determine if we had chosen the proper facts for the model (this step eliminated the false offset mapping shown in Fig. 6-7).

Now that we have the feel of a more complex problem, let's try one that takes us a little farther out.

Fencing the Pasture

This time, let's imagine that you have one sheep to put into a pasture and 88 feet of fencing to use for enclosing it. You want to fence in as much ground as you can with your 88 feet of wire, so that the sheep will have as much grazing room as possible.

The problem, then, is that of determining the optimum shape and size to make the pasture.

We could start with the conventional rectangular shape. The distance around a rectangle is equal to the sum of all the sides, and the opposite sides are equal length. Thus, the distance around a rectangle 2 feet wide and 10 feet long would be 2 + 10 + 2 + 10 feet, or 2 times (2 + 10) feet, which is 2 times 12 or 24 feet.

If we're going to use the rectangular shape, our problem is to choose a length and width which will come out to a total perimeter of 88 feet. We can map this as follows:

1. perimeter = 2 × (length + width)
2. desired perimeter = 88
3. 88 = 2 × (length + width)
4. dividing each side by 2, 44 = (length + width)

The final equation of this sequence shows us that in a rectangle, with a fixed perimeter of 88 feet, the sum of length and width must equal 44 feet. The precise length is determined by the width we choose.

That is, if we treat width as an independent variable in the range 0 to 44 feet, the length becomes a dependent variable. This dependency is expressed by the following sequence:

1. length + width = 44
2. length + width − width = 44 − width
3. length = 44 − width

If we now substitute any value in the range 0 to 44 feet for the width, the final equation shows us how to determine the corresponding length.

For a width of 1 foot, the length is 43 feet. If we double the width to 2 feet, the length becomes 44−2 or 42 feet. Doubling the width again to 4 feet, the length drops to 44−4 or 40 feet. If we increase width to 8 feet, the length is 36 feet. At a width of 16 feet, the length is 28 feet. At 20 feet wide, the length is 24 feet. When the width rises to 24 feet, the length drops to 20. In each case, the length and width added together total out to 44 feet, so the 88 feet of fencing would be completely used.

The problem was not just to use all the wire, though. The idea was to do it in such a way as to get the largest possible pasture for the sheep to graze in. This requires that we use the shape and size that produce maximum area.

Area is determined by multiplying length times width. Let's see how the area of the various sizes we have suggested compares, one to another:

WIDTH	LENGTH	AREA
1	43	1×43 or 43
2	42	2×42 or 84
4	40	4×40 or 160
8	36	8×36 or 288
16	28	16×28 or 448
20	24	20×24 or 480
24	20	24×20 or 480

If we keep going, we will find that a 28-foot wide by 16-foot long rectangle has the same area as one 16 feet wide by 28 feet long. This indicates that the area reaches a maximum between the 20×24 size and the 24×20 one.

If we tabulate sizes in this region, we will come up with something like this:

WIDTH	LENGTH	AREA
20	24	480
21	23	483
22	22	484
23	21	483
24	20	480

It appears that of the square-cornered shapes, a 22-foot square will give the largest area. If this is true, we should be able to prove it logically; so let's give it a try.

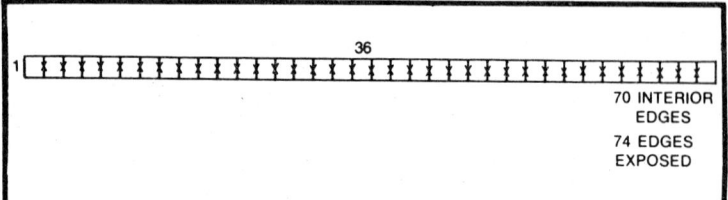

Fig. 6-9. If we have 36 square tiles and put them all in a row, 74 of the total of 144 tile edges will be exposed and only 70 will be on the inside of the area covered. This shows that the long, narrow shape has a large perimeter for its area.

To make this proof, let's consider a group of 36 square tiles. They may be arranged in any rectangular pattern: one tile wide by 36 tiles long, 2 wide by 18 long, 3 wide by 12 long, 4 wide by 9 long, 6 wide by 6 long, 9 wide by 4 long, 12 × 3, 18 × 2 or 36 × 1.

Since in each of these patterns we used the same 36 tiles, the area was the same in every case. When we made the pattern one wide by 36 long, or vice versa, two sides of each of 34 of the tiles, and three sides of each of the other two, were on the outside edges of the rectangle contributing to the perimeter (Fig. 6-9). Of the total 144 edges on all tiles together, only 70 (or less than half) were not exposed to the outside of the figure.

When we form a 2 × 18 or 18 × 2 figure, the situation changes. Only two sides of each of the corner tiles and one side of each of the others, is on the perimeter (Fig. 6-10). Three sides of 32 tiles, and two sides of the other 4, are interior edges, so 104 edges are not exposed.

In the 3 × 12 figure (Fig. 6-11) it's the same—except that now 10 of the 36 tiles are completely inside the figure and contribute no edges to the perimeter. Of the 144 edges available, 114 are not exposed.

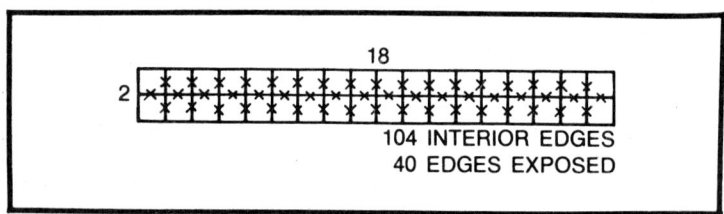

Fig. 6-10. Changing the shape of our 36-tile rectangle to 2×18 gives us only 40 edges exposed, a reduction of 34. We can do better, though, at reducing the perimeter while keeping the area.

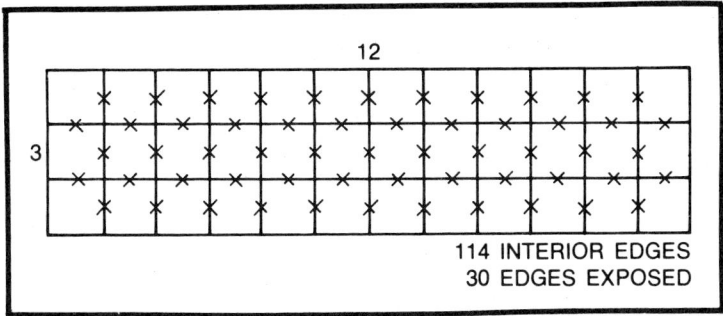

Fig. 6-11. The 3×12 shape reduces the number of exposed edges to 30, but we still do not have the minimum perimeter for the 36-tile area.

The 4 × 9 pair of figures continues the trend. Here, 14 tiles are totally interior, so that the number of interior edges rises to 118.

When we reach the 6 × 6 square pattern (Fig. 6-12) only 20 of the 36 tiles are on the edges, and only 4 of these 20 have more than one edge exposed. The number of interior edges is 120.

That is to say, the square shape puts more edges on the inside than a rectangle would, so that for a given area, the perimeter will be smallest for a square.

This analysis assumed that we held the area constant and let the perimeter change; and it shows that if we do so, the square will have the smallest perimeter for a given area.

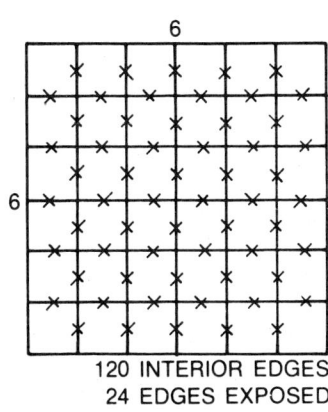

Fig. 6-12. A 6×6 square exposes only 24 edges of the 144 which are present. Thus the square gives the shortest perimeter for a given area, which is the same thing as saying the largest area for a given perimeter, of any rectangle. Have we reached the ultimate yet?

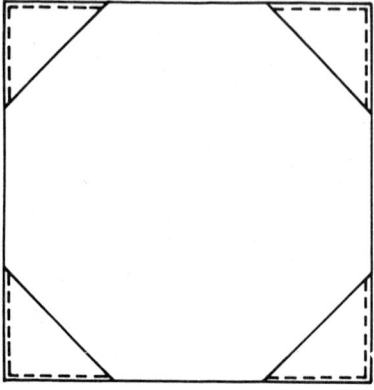

Fig. 6-13. If we cut the corners off the square, we will get more area for the perimeter. Or will we?

If we do it the other way around, holding the perimeter constant and letting the area change, we will have to reduce the area for any shape but the square to reach that smallest perimeter value; so this proves that for a fixed perimeter length, the square has the greatest area of any square-cornered figure.

This tells us that we can get the most area for a square-cornered pasture, with our 88 feet of fencing, if we make it in the form of a square 22 feet on a side.

But it does not prove that we have the biggest pasture we can get for our 88 feet of fencing. What if we forget about the square-cornered shape? What happens then?

For instance, we can clip off the corners of our square to form an octagon (8-sided figure) as shown in Fig. 6-13. With 88 feet of fencing, we could make each of the 8 sides 11 feet long. What would be the area of the resulting pasture?

We know that the area of a rectangle is equal to its length times its width. That of a right triangle is half as much as if the corresponding triangle making up the other half of the rectangle were present, or half the length times the width. To calculate the area of our octagon, we slice it up into a square, four other rectangles, and four triangles, as shown in Fig. 6-14. Now we can calculate the area of each individually, then add them together to get the area of the entire pasture.

The area of the square is easy. It's 11 feet on a side, so its area is 11 × 11 or 121 square feet.

To get the area of any one of the other four rectangles (all are the same size, so we can figure the area of one and then use it four times), we have to work Pythagoras' theorem to determine the length of the short side. This is also the length of the sides of the triangles.

We'll do the work on one of the triangles. The symbols are shown in Fig. 6-14. We're using the conventional A, B, and C symbols, where C is the 11-foot hypotenuse, and A and B are equal.

If $A = B$, then the Pythagorean equation $A^2 + B^2 = C^2$ can also be written as $2A^2 = C^2$, and this can then be rearranged into $A^2 = C^2/2$. When we then take the square root of both sides of the equation, we find that $A = C/\sqrt{2}$.

We'll temporarily represent the square root of two by the symbol R (for root) to simplify the printer's job. The length of sides A and B is, from our last equation, $11/R$. Therefore, the area of each of the four triangles would be half of $(11/R) \times (11/R)$. The quantity $(11/R) \times (11/R)$ multiplies out to $(11 \times 11)/(RR)$, which is $121/2$. Half of this is $121/4$. Since we have four triangles, we can multiply by 4 to get 121 square feet as the total area of all four triangles. This brings our known area up to 242 square feet, with the four rectangles still to go.

The area of each of them is $11/R \times 11$, or $121/R$ square feet. Since there are four of them, the total area is $(4 \times 121)/R$. This can be approximated for our calculations in either of two ways.

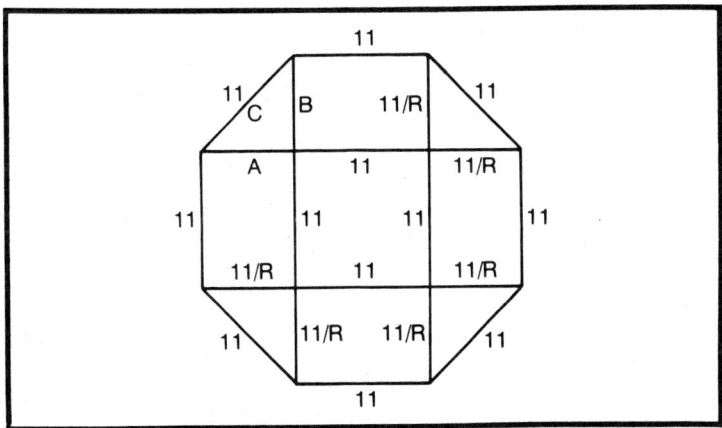

Fig. 6-14. Here's how we divide up the octagon to calculate its area. Each side is 11 ft, to use the 88 ft of fencing. In the middle we have an 11 ft square. It is bounded by four rectangles, each 11 ft on one side and shorter on the other, and four triangles, which form two squares, each the same size as the short sides of the rectangles. Text works it all out.

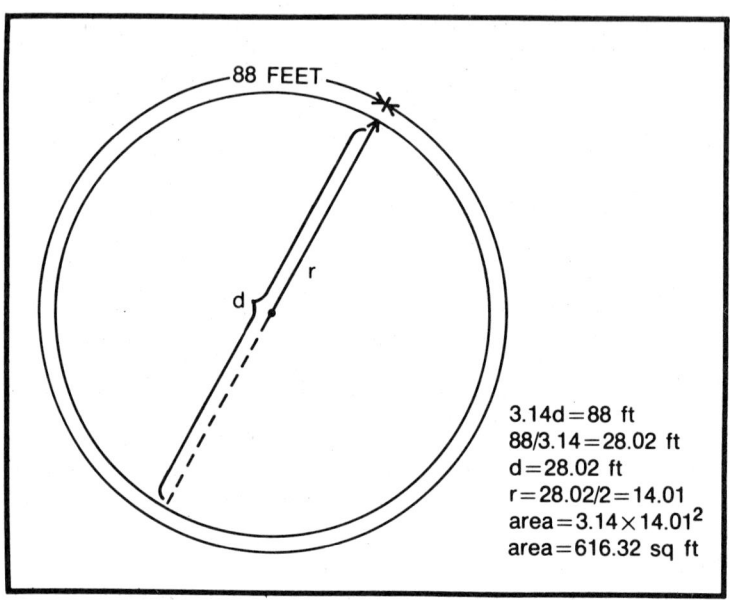

Fig. 6-15. When we get down to it, the circle gives the largest area for a given perimeter. Area is almost half again larger than corresponding square.

The most direct method is to multiply out 121 × 4 to get 484, then divide by the approximate value of R (1.414).

The more elegant method, which changes the arithmetic from division to multiplication, is to factor 4 into 2 × 2, and then factor one 2 into $R \times R$, so that the 4 becomes $RR2$ or $2R^2$, and the 4 × 121 becomes $R \times R \times 2 \times 121$ or $R \times R \times 242$.

The R below the bar then cancels one R above the bar, leaving the total area as $R \times 242$ square feet. Multiplying this, using 1.414 for R, gives us 342.188 square feet for the area of the four smaller rectangles.

Adding the 342.188 figure to the 242 square feet we already had calculated brings the total area of the octagon to 584.188 square feet.

This is nearly 100 square feet more than we could get with the square, which was the best a square-cornered figure could provide. Can we get more?

If we keep clipping off the corners, we will eventually get to the circle shown in Fig. 6-15. (Actually, we could never form a circle by clipping corners off a regular polygon any more than we could form a real number by using enough digits in a rational one—but, as in numbers, we can get a workable approximation.) The area of a circle is equal to pi (3.14159) times the radius squared, while the perimeter or circumference is equal to pi times the diameter. For an 88-foot circumference, using the approximation 3.14 for pi, we find that we can form a circle 28.02 feet in diameter. The 14-foot radius means that the approximate area is 616 square feet.

And this is the best we can do. We have no more corners left to clip. Our sheep, fenced into a 28-foot circle, can graze on more than 600 square feet of pasture.

This exercise in mathematical model-building took us quite a bit farther than it might appear on the surface. As in the previous example, we had to set up the problem and work through some false starts (which actually helped give us insight into the real problem involved), but here we also established a mathematical fact which pops up unexpectedly many times—that a circle has the largest area for its perimeter of any figure on the plane.

The method we hinted at of "trimming off the corners" is actually the way the accuracy of the formula for calculating area of a circle is proved. You start with a square outside the circle, and another inside, and "trim off the corners" repeatedly until it's apparent that you're getting a closer and closer fit (a la the cut which establishes a real number between adjacent rationals). The resulting pair of series calculations for area of the two figures turns out to approach the series for pi.

Now let's try a gambler's problem, which will show the difference between a professional gambler and the suckers who keep him in business.

The Chuckaluck Game

One of the traditional gambling-machine games found at touring carnivals is chuckaluck. It's a basket shaped somewhat like an hourglass, and it contains three dice.

The dice are conventional. Each face of each die is numbered by a pattern of spots, from one to six.

The players can choose any number from 1 to 6. When all the bets are down, the operator of the game spins the basket to tumble the dice.

If a player's number appears on any one of the three dice, he is paid the amount of his bet. That is, if it cost $1 to play, he gets $2—his bet returned, plus that much more from the operator.

If the number appears on two of the three dice, he gets paid double; for a $1 bet, he gets back $3.

And if his number appears on all three dice, he is paid triple—$4 for a $1 bet.

At first glance, this looks like a game at which the operator is bound to go broke in a hurry. In fact, however, he can't escape making money, and the bigger the payoffs he makes, the more he pockets.

The problem we're going to try to solve here is not *Why is this so?*—we're going to have to find that out before we can begin working on the problem itself—but rather *What payoffs would be necessary to turn it into a break-even game?*

Along the way, we're going to find out something about probability and the laws of chance.

First, let's see how the operator of a chuckaluck game can make so much profit out of paying out on a 1-2-3 basis. Let's assume that he has six players, and that each of them has put his money on a different number.

When the dice stop rolling, all three dice must be showing numbers in the range 1 to 6, and we have assumed that each number has a player who has put $1 on the line to play the game.

One of three situations must be the case when the dice stop: either each die shows a different number, two dice show the same number and the third is different, or all three show the same.

No matter which happens, the operator has six $1 bets in front of him, which add up to $6.

If each die shows a different number, the game has three winners and three losers. The operator gives each winner the extra $1 from one of the losers. The operator makes no profit, but loses nothing either.

If two dice show the same and one is different, there are two winners and four losers. One of the winners, the single-die one, gets paid off with the bet of one of the losers. The other winner gets his $2 payoff from the bets of two more of the losers. The fourth loser's dollar goes into the operator's pocket, this time there was a profit.

If all three show the same, there is one winner. The other five are losers. The winner gets his $3 payoff out of the $5 lost, and the operator pockets the other $2.

Thus, the bigger the payoff, the more the operator makes.

Of course, if he doesn't have the same number of players every time it isn't quite so direct in practice—but in the long run it works out just the same, because the bias in favor of the operator is built into the payoff odds.

In order to find out what the payoff should be to keep the operator from making a profit, we're going to find out just how honest gambling odds are figured and used.

Let's look first at the situation with a single die. The conventional gambling die is a cube (Fig. 6-16) with spots on each face. The spots on opposite faces total a count of 7, so that the number of spots on each face ranges from 1 to 6. No

Fig. 6-16. Conventional dice are used in game of chuckaluck. Three of them are spun in a basket. Operator make his biggest profits when he is paying off largest winner.

number is duplicated on an honest die, so that the result when it is rolled must be a count from 1 to 6.

There are 6 possibilities, and for each roll only one outcome. The chance of any specific number coming up, then, is one in six; we say that with a single die the probability of rolling any specific number is 1/6.

Another way of expressing the same fact is to note that there's only *one* way for the number to come up and *five* ways for it not to. We can then say that the odds are five to one against rolling any specific number.

The odds indicate what a no-profit payoff would be. Imagine that you like, for some reason, the number 4 and keep betting a dollar on it for six consecutive rolls of the die. Imagine, while we're at it, that each of the six possible numbers comes up exactly one time in the sequence of six rolls.

You've paid $6 for the six rolls at $1 each. At chuckaluck odds, you got back $2 when the 4 came up, but nothing at all the other five times. You lost $5 to win $1, for a net loss of $4. Your $4 loss is a $4 profit for the operator.

If the operator paid on the mathematical odds, he would pay $5 rather than $1 for the single-die win. In this case, you would pay $6 for the six rolls, but get back $6 when the 4 came up. Neither you nor the operator would win or lose over the series; it would be a no-profit game for both of you with only one die in use.

When we add the second die, things change. Now instead of 6 possible outcomes, there are 6 outcomes for each of the two dice, and each is independent of the other. This gives us 6×6 or 36 possible outcomes for the pair together.

As before, if you choose a single number to come up on both, you have only one way to win the double which means there are 35 ways you can lose it. There are now, though, two times 5, or 10 ways to win singles.

The mathematical odds on the double are 35 to 1 against your winning. The probability of winning it is 1/36.

For a no-profit payoff, the game operator should pay you $35 instead of $2. The fact that the 35 ways to lose the double include 10 ways to win the single modifies the odds somewhat.

Putting in the third die makes another change in odds. Each of the 36 outcomes for two dice gets 6 more from the third, giving $6 \times 6 \times 6$ or 216 possible outcomes, and there's only one winning triple combination for any player. That means there are 215 ways to miss the triple payoff.

Mathematical odds in this case are 215 to 1 against, and probability is 1/216. For a no-profit payoff, you should get $215 rather than $3, except that the 215 ways to lose the *triple* include the ways to win on *double* and *single* odds.

It appears that the chuckaluck operator has a sure thing going for him, with a rather good margin of profit.

It may not be so apparent that he will have just as sure thing going if he increases his payoffs, so long as he always pays off at odds lower than the true figures.

The key difference is that the traditional odds are set so low as to require little or no capital on the part of the operator. Most honest odds mean that—in case of a run of bad luck—the operator may require a sizable bankroll.

However, shaved odds are precisely why professional gambling establishments such as those in Nevada and Monaco stay in business. They do not cheat, but the odds are always shaved slightly in favor of the house.

For instance, the game of roulette involves a ball that can fall into one of 38 slots in a wheel. The game pays off for any number on 35-to-1 odds.

Similarly, payoffs on all other games are modified to provide a house profit in the odds. Except for this, the action is scrupulously honest—yet it's a mathematical certainty that in the long run, the house will win, because it's built into the system.

If the laws of probability and chance dealt only with gambling, they might not have much interest for many folk. However, they also form an important part of modern science. For instance, we noted that the occurrence of a 4 on one die in the chuckaluck game had a probability of 1/6, while the probability of 4 on each of two dice at the same time was 1/36. This is 1/6 times 1/6; if we know the probability of each of two events separately, we can find that of the two happening together by multiplying the individual probability figures.

If we're satisfied with either a 2 or a 4 on a single die, we have two ways to win and four to lose, for a probability of winning of 2/6 or 1/3. This is the same as 1/6 plus 1/6. We can find the probability of *either* of two events by adding the individual probability figures.

The same addition and multiplication of probability figures is used to deal with problems in nuclear physics, to validate tests of medical techniques, and in many other branches of science. It helps establish the rate you pay for various kinds of insurance, and makes it possible for you to buy insurance in the first place.

The applications to insurance bring out the differences and similarities in *a priori* and *empirical* probability. The dice we have been discussing exhibit *a priori* probabilities; that is, the total number of possible outcomes is fixed by the design of things, prior to the first trial toss.

Insurance firms, on the other hand, deal with *empirical* probabilities. They may know that the probability of a home being destroyed by fire in a certain city is 1/319, and so be willing to insure against that destruction on the basis of 318-to-1 odds against the event—but they have no list of possible outcomes to let them calculate this probability. What they have, instead, is a *claims record*, which lists how many losses they have had in the past several years in that city, and they know that for every 319 homes in the city, one has been destroyed by fire in the preceding year.

That is, they keep track of what has actually happened, and assume that the same thing is likely to happen again. One of the reasons for all the small print in an insurance policy is to make certain that this assumption is valid. Outbreak of war, for instance, throws the records out as a valid indicator of probability of loss—so they exclude war losses from their coverage.

The only difference between the two kinds of probability, though, is how they are originally arrived at. Once you get a probability figure in the range 0 (cannot happen) to 1 (certain to happen), you use it like any other.

Chapter 7

Puzzling it Out

Most of the real fun of mathematics (and by now, if you're still with us, you must believe that it can be fun) comes from its "solve the puzzle" aspects. The only usefulness that math can have in the real world is that it provides us with a tool we can use to solve problems of one sort or another. If we're not using it to solve a problem which faces us in the real world, then we're simply figuring out the answer to a puzzle.

Professional mathematicians find great fascination in such puzzles as the question *Is it possible to predict whether any given number is prime?* but for most of us, puzzles of this sort are a bit too deep to be much fun.

Even the simplest everyday puzzle, though, is an application of math when it involves the use of logic. Let's see how it works with some problems which are a little bit closer to the real world than they are to the world of the professional mathematician.

Take, for instance, the problem faced by a farmer who is going to the marketplace to sell a fox, a hen, and a bushel of corn. One of his problems is that the fox, given a chance, will kill and eat the chicken; the hen, left alone with the corn, will consider it chickenfeed and do her best to reduce the farmer's burden. The only way the farmer is going to make it to market with all three intact is to be certain he does not leave the fox alone with the hen, or the hen alone with the corn.

During the early part of the farmer's journey, this is no special trouble. However, between his farm and the marketplace is a river, and when he reaches the river he finds that the only way to get across is in a small boat boat which has room for only himself and one of his burdens. He must leave two unattended while he takes the third across.

If he takes the corn across and leaves the fox and hen behind, the fox will eat the hen. If he takes the fox across and leaves the hen with the corn, the corn will disappear. The same problems occur if he takes the hen across first and leaves her on the far side: No matter which he takes across on his second trip, he's going to find trouble after he leaves the hen unattended with either of the others.

Yet the canny farmer managed to solve his problem and get all three across the river safely, never taking more than one at a time in the boat with him. The puzzle is, how did he do it?

We can map the problem out by using symbols to stand for the individuals involved; let's let M (for man) represent the farmer, F the fox, H the hen, and C the corn. When the group reaches the river, we have M, F, H, and C on the west bank; the problem is to get all four to the east bank intact. (We'll assume that neither the hen nor the fox will run away if left alone.)

We can represent the fox's chicken dinner by the expression $FH = F$, meaning that if the fox and the hen are left alone together only the fox will remain, and similarly $HC = H$ can be used to mean that leaving the hen with the corn will eliminate the corn.

Now let's start ferrying things across the stream. We start with $MFHC$ on the west bank. If the man takes the corn on his first trip, M and C are removed from the west-bank group, leaving only FH. But $FH = H$, so when he returns he finds only F on the west bank. This is not a successful outcome, so he cannot take the corn first.

If he takes the fox first, the west-bank group is reduced to HC, and since $HC = H$, this outcome is also unacceptable. Therefore, he must take the hen across on the first trip, leaving the west-bank group as FC, which will not result in damage since the fox will not eat the corn.

Having taken the hen across, the farmer returns for another trip. On the west bank is FC, and on the east bank now is H.

Either the fox or the corn can be taken across on the second trip, since only one item would be left on the west bank and no damage can result.

But no matter which he takes across the second time, the farmer cannot leave the hen on the east bank on his second return trip. If he took the corn, leaving the hen would result in $HC = H$; if he took the fox, it would result in $FH = F$. Thus, the farmer must bring the hen back from the east bank to the west the second time he returns. While he is making the crossing, the group on one bank is F and that on the other bank is C; each is alone, so no damage can result.

Having brought the hen back to the west bank, the farmer must leave her there while he takes the last item across. After this crossing, the west bank is H and the east bank is FC.

On the final trip, the farmer takes the hen across again, and then all three of his charges are on the east bank. The problem is solved.

We can show the sequence of crossings in a table:

	WEST BANK	CROSSING	EAST BANK
1.	F H C M		----
		H M	
2.	F C		H M
		M	
3.	F (or C)		H
		M C (or F)	
4.	F (or C)		C (or F)
		M H	
5.	H		C (or F)
		M F (or C)	
6.	H		F C
		M	
7.	---		F C
		M H	
8.	---		F H C M

The logic involved here is simply to keep in mind that at no time during the entire action can the set on either bank consist of *FH* or *HC*. Any sequence that leads to either combination is a disastrous outcome.

TRUTH-TELLERS

Another class of logical puzzle involves people who always tell the truth and those who always lie.

One of the more mind-boggling of these puzzles is a simple card. On one side is written the sentence *The statement on the other side of this card is true*; on the other side is written *The statement on the other side of this card is false*.

Assuming that you have no other information available, how can you tell which side of the card is correct?

If you decide that the first is correct, it asserts that the statement on the other side is true also—but *that* one tells you your decision was wrong.

If you decide that the first was wrong, it would imply that the statement on the other side was not true—and that conclusion would then tell you that your decision was wrong in this case also.

One of the rules of the game in mathematical logic is that you cannot draw a meaningful logical inference from a false statement. Therefore, if we decide that the first statement was wrong, we really have no information at all about the truth of the second statement. But this leads us to the same problem. If the second statement is true, that means that the first had to be false—and that one claims that the second was true. The pair of statements has a built-in contradiction. If either of them is true, it must at the same time be false, and this is not allowable.

Such a built-in contradiction is called a *paradox*, and professional mathematicians find great pleasure in attempting to find loopholes in logic which make paradoxes possible to state (since they obviously are not possible in the real world, their existence reveals shortcomings of the most basic tools of math).

But until some way is found to account for paradoxes, the answer to our puzzle is simply that neither side can be correct if the other is. Not very satisfactory to a logical mind, but it's the only answer available.

Less confusing, and therefore more enjoyable to most of us, are the nonparadoxical puzzles involving liars and truth-tellers. For instance, there's the problem faced by the missionary who landed on a South Sea island inhabited by two tribes. One tribe lived on the north side of the island and the other on the south. They looked alike, dressed alike, and acted the same way. The only way to tell them apart was by what they said, since those who lived in the south always told the truth, while those from the northern side of the island never did. The missionary wanted a native assistant in his work; and quite naturally he wanted one who always told the truth.

The problem was how to test a candidate with only one question, to determine to which tribe he belonged—and the missionary figured out a way.

He told his selected candidate to ask a second native in the distance which side of the island he lived on. The candidate did as he was asked, and returned with the answer, "He says he lives in the south."

Was the candidate a truth-teller or not, and how did the missionary know?

If the second native lived on the south side of the island, he would have said so when asked, since southerners are always truthful.

If, on the other hand, he lived in the north, he would have said he lived on the south side, since the northerners always lie.

Therefore, no matter which tribe the second native belonged to, he would reply to the question *Which side of the island do you live on?* with the answer, *the south.*

Since the candidate reported this as being the answer, he must be a truthful southerner.

Had he replied, "He says he lives in the north," the candidate would have revealed himself as a northerner. Thus, by making the answer to the first question completely independent of the truth or falsity of the answer, the missionary devised a manner of testing the truthfulness of his would-be assistant.

Similar principles apply to many other puzzles. Another Polynesian island, so the story goes, is inhabited by equal numbers of brown-skinned natives and immigrant whites who have become sunburned to the same shade of brown.

On this island, the white men always lie and the natives always tell the truth. One day a visitor to the island saw three inhabitants on the beach and called out to one of them "Are you a native or a white man?"

The individual's reply was drowned out by the crashing surf, and so the visitor called out the same question again. This time one of the others replied, saying "The first man said he was a native, and he *is* a native, and so am I."

The third person interrupted to shout, "The first two who answered are really whites burned brown, but I am a native."

Can you tell from this which of the three men is white and which is native?

Like the second native in the puzzle of the missionary's assistant, who said the same thing whether he was truthful or lying, the first man to answer must have said he was a native. If he was a native, being truthful, he would say so. If not, being a liar, he would claim he was.

Since the second person said that the first claimed to be a native, he had to be telling the truth. Since on this island anyone who ever tells the truth always does so, he was truthful when he said the first man was a native; therefore the first two persons must be natives.

The third man's claim that the first two were really whites must then be false, which establishes that he himself must be a sunburned white.

Still another problem in this family concerns a traveler in hillbilly country who comes to a fork in an unfamiliar road. The traveler wants to get to Nashville, but doesn't know which fork to take.

Two men are standing nearby; one of them is truthful and the other is a consistent liar. The traveler has no way of knowing which is which, but he asks one man one question and learns the right way to Nashville from the reply.

What question did the traveler ask?

We've already seen how it's possible to set up a question so that it must have the same answer whether the person who answers tells the truth or lies, but that's not the exact problem we're facing here. We don't really care whether one man or the other is telling the truth—we must find out how to determine the correct way to Nashville regardless of the truth or falsity of the answer.

The basic principle involved is exactly the same, but its application differs. To frame a question which has the same answer whether it's true or false, we must phrase it so that the truth-teller's answer is forced from the consistent liar.

The missionary used this to determine whether his candidate was lying or not; the traveler can put a twist into the technique to learn the right route to Nashville.

He cannot ask *Which is the way to Nashville?* because if he is answered by the liar, he will get the wrong reply.

What he *can* do is to ask a question involving the answer to another question, so that the two lies cancel out if he is talking to the liar. He asks either man, "If I were to ask you if this fork leads to Nashville, would you say yes?"

If the liar replies and says "Yes," then he is in effect saying that if the real question had been *Does this fork lead to Nashville?* his (untruthful) answer would have been *no*. That, in turn, in effect says that this fork does actually lead to Nashville.

If the liar replies and says *no*, then he is in effect saying that his answer to *Does this fork lead to Nashville?* would be *yes*, and since that too would have been a false statement, this fork does not actually lead to Nashville.

If the truthful man replies with a *yes*, he is saying that his reply to *Does this fork lead to Nashville?* would be *yes*, and that would be the case.

Finally, if the truthful man replied with a *no*, he is saying that his reply to *Does this fork lead to Nashville?* would be *no*; since he is truthful, this fork does not lead to Nashville.

By tallying up the four possible outcomes of the question, as we just did, we can see that if the fork really does lead to Nashville, both the liar and the truth-teller will reply with a *yes* to the actual question asked, and that if the answer is *no*, it makes no difference which of them said it.

In effect, the question-within-question makes the liar lie twice so that the two cancel each other out, and that's one of the basic principles of mathematical logic—two inversions (from true to not-true to *not* not true) cancel.

Unfortunately, the principles involved in these puzzles don't help much in the real world, where hardly anyone is totally consistent in either truth or falsehood. The realm of logic is far removed from reality, even though many things in everyday life seem to have some logical basis. The difference is the consistency; logic is all-or-nothing, while reality is always somewhere between.

CHECKING UP ON THINGS

Another whole class of logical problems has to do with the classification of items into sets, as we did in a previous chapter. Sometimes the actual problem of the puzzle involves finding out how such classification can be done.

Fig. 7-1. This is the starting point for the four-card puzzle. How many cards must be turned over to prove or disprove the claim that every red card has a circle on its other side?

Consider the four pieces of cardboard shown in Fig. 7-1. Each of them is either red or green on one side, and has either a circle or a cross on the other.

At the start of the puzzle, they are lying on a table just as shown in Fig. 7-1, with one red side showing, one green side, one circle, and one cross.

The problem is simply to determine how many (and which) cards must be turned over in order to be able to prove or disprove the claim that *every red card has a circle on its back*.

From the starting point, there are only 16 possible combinations of the cards. If we represent each card by a pair of symbols, using R and G to indicate red or green sides and O and X to indicate circles and crosses, we can represent a card which is red on one side and has a circle on the other by the symbol pair RO. The 16 possible combinations of cards are listed in Table 7-1.

Had we not been given the starting point, there would have been 256 possible combinations, since any card could have been either an RO, RX, GO, or GX, and with four cards and four possibilities for each we could have 4^4 (16^2 or 256) patterns. The first card being red eliminated all those for which the first card was green, or half of them. The second card being green eliminated half of what was left. The third card having a circle took out half of the remainder, and the cross on the last cut the total to 16.

If the claim that "every red card has a circle on its back" is true, then the RX pattern cannot exist. If we find a single case of the RX pair, we have disproved the claim.

174

Table 7-1. Card Combinations

	First Card	Second Card	Third Card	Fourth Card
1	RO	GO	RO	RX
2	RO	GO	RO	GX
3	RO	GO	GO	RX
4	RO	GO	GO	GX
5	RO	GX	RO	RX
6	RO	GX	RO	GX
7	RO	GX	GO	RX
8	RO	GX	GO	GX
9	RX	GO	RO	RX
10	RX	GO	RO	GX
11	RX	GO	GO	RX
12	RX	GO	GO	GX
13	RX	GX	RO	RX
14	RX	GX	RO	GX
15	RX	GX	GO	RX
16	RX	GX	GO	GX

Table 7-1 shows us the distribution of the possible patterns. The *RX* pattern shows up on lines 1, 3, 5, and 7, as well as 9 through 16.

But every time the *RX* occurs, it is either on the first card or the fourth card. Therefore, we need turn over only two cards (the first and the fourth) to prove or disprove the claim. If the red card has a circle on its back, the claim is unproven and we have to look at card number four, which has a cross showing. It may be either red or green on the other side; if it's green, then we know that every red card has a circle on its back and the claim is proved.

Most people who meet this puzzle begin by assuming that it's necessary to look at card number three, which shows a circle, to see whether it's red or green on the other side. The list of possible patterns shows that this doesn't enter into the proof or disproof of the claim, however. Only two cards need be checked to prove it, and disproving it could happen if the first card checked failed to satisfy the claim.

In this same general family of puzzles is one which true puzzle fans probably already know, but which involves two levels of logic (somewhat like the double-negation trick we met in the case of the truth-tellers) and therefore fits into our scheme of things right here.

A corporation executive was in search of an assistant, and he needed one with the ability to rapidly reach logical conclusions. After narrowing his field of applicants to three individuals who seemed equally suited for the job, the executive had to devise a test to determine which of them could reach valid conclusions most rapidly.

He did so by having each of the three close his eyes while the executive placed a mark on his forehead. The mark was either red or black. Every man was marked. Then, on the executive's command, all three opened their eyes and looked around them. As soon as anyone saw a red mark, he was to stand up. As soon as he could tell what color his own mark was, he was to sit down. The first candidate to sit back down would get the job.

When the executive gave the word, all three opened their eyes, and almost immediately all three stood up.

A moment later, one of the three sat down.

What color was his mark, and how did he figure it out? You have all the information he had—and it's strictly logical, but it involves two levels of logic.

When the successful applicant opened his eyes, he saw two red marks on the other two. As he stood up, so did the other two—but either one of them could have been responding to the red mark on the other's forehead, so this by itself gave the successful applicant no information about the color of his own mark.

However, as time passed and neither of the others sat down, he realized that they must be as uncertain as he was about the color of their own marks.

If either of them could see only one red mark, he could determine from that and the fact that all three were standing that his own mark had to be red—because if his own were black, the one man with a red mark would not have stood up.

Since both were still standing, they were both still undecided, and this meant that each of them must be seeing two red marks.

Therefore all three marks must be red, reasoned the successful applicant. When he reached this conclusion, he sat down and the job was his.

The puzzle can be extended to four applicants, provided that each of them is familiar with the three-person puzzle and the ground rules are changed so that anyone who sees *two* red marks must stand.

In this case, the successful applicant is the one who realizes that if his own mark were a different color than the three he can see, then the problem would become that of the three-person puzzle for the other three, and one of them could solve it (since all four are familiar with it).

Since nobody has solved it, then his mark must be the same color as those which he sees, and he wins.

It might appear that we could keep on extending this sort of thing indefinitely—but we can't. If we try it with five people, there is no guarantee that the others would have been able to solve the four-person version.

WHO'S ON FIRST

Still another category of logical problems involves the very roots of mathematical logic. In this class of puzzle, which enjoys periodic surges of popularity, you are given a string of facts which appear to have little or no connection with each other, and asked a question which is not obviously related to any of them.

The problem is to answer the question, working only from the given facts.

For example, consider three men named Alvin, Bill, and Charley, who live in adjacent houses on the same street. Each of the three has a pet, and no two of them have the same kind of animal. Each house is painted a different color from the other two. Alvin's house is white. Charley owns a monkey. In the red house lives a cat.

The question is: *What color is Bill's house?*

This is one of the simplest examples of this kind of puzzle, since it requires only two deductions.

Alvin cannot own the cat, since his house is white and the cat lives in the red house. Neither can Charley, because Charley owns a monkey. The cat, then, must live with Bill in the red house, so Bill's house is red.

The basis for this whole class of logic puzzle is the theory of sets, which we met back in Chapter 2. We can set up the sets of people, pets, and house colors as *Alvin, Bill, Charley; monkey, cat, something else;* and *white, red, other color*. Now we try to map these sets together from the partial mappings given us by the set of given facts: *Alvin, ??, white; Bill, ??, ??;* and *Charley, monkey, ??*.

The other partial mapping given in the original set of facts is *??, cat, red;* and the only place this mapping can fit in is to match *Bill, ??, ??;* this gives us *Bill, cat, red* and answers the question asked by the puzzle. We can then complete the mappings just to clean things up by fitting the "something else" and "other color" we had left over into the only remaining vacancies. This gives us the final mapping: *Alvin, something else, white; Bill, cat, red;* and *Charley, monkey, other color.*

If you would like to design other puzzles of this sort for yourself, all you need do is reverse the process. Take any set of facts that can be expressed in a form similar to our final mapping, and then replace some of the entries in each set of the mapping by question marks (as in our partial mappings along the way). Finally, phrase the remaining parts of the partial mappings as statements of fact, and ask a question involving two of the parts which you removed.

The more different sets of facts you use in the mapping, the harder will be the puzzle to solve (if you do it right).

Here's one with four rather than three sets:

Abner drives a Chevy. Bob went hunting. Carl's vehicle is a Jeep. Dave slept later than Carl. The Ford driver got up first. Abner slept longest. The man who drives a Plymouth went swimming. The second man to get up went fishing.

If no two of the men drive the same kind of car and all did different things, who played golf?

We can start to solve this one by arranging the given facts into partial mappings following the pattern *name, car make, sport, sequence of rising*. This gives us:

Abner, Chevy, ??, fourth; Bob, ??, hunting, ??; Carl, Jeep, ??, ??; and *Dave, ??, ??, ??.*

In addition, we have the following facts left over:

??, Ford, ??, first; ??, Plymouth, swimming, ??; ??, ??, fishing, second; and (from the question) *??, ??, golf, ??.*

We can match Bob's set and the Ford driver's, and cannot match any other set with that for the Ford. Similarly, we can match up Carl and the fisherman. When we do this, we have:

Abner, Chevy, ??, fourth; Bob, Ford, hunting, first; Carl, Jeep, fishing, second; and *Dave, ??, ??, ??.*

Still left over are the facts *??, Plymouth, swimming, ??;* and the question *??, ??, golf, ??.*

Since every man's car is identified except for Dave, he must be the Plymouth driver. This gives us *Dave, Plymouth, swimming,* ??; and leaves Abner as the only possibility to be the golfer.

Notice that the fact *Dave slept later than Carl,* while true, did not enter into the solution. Many such puzzles have extra facts of this sort tossed into the list to help make them more puzzling.

No matter how many facts are given, the technique of making partial mapping will lead to a solution. With a well-constructed puzzle, however, the mappings will not be as simple to create as were the two in our examples.

NUMBER PUZZLES

The most obviously mathematical class of puzzles is that group which involves numbers.

For instance, there's the problem faced by the camp cook who found it necessary to measure four ounces of vinegar from a jug, but had only a five-ounce and a three-ounce container, neither of which was marked off in ounces. How did he do it?

This one involves a whole series of logical steps. The only way the cook can get four ounces measured into the containers is to find a method for taking one ounce out of the five-ounce one, since the three-ounce one cannot hold all four ounces.

He might try to find a way of measuring one ounce into the five-ounce container, so that he could then add the three ounces from the other container to get four—but within the ground rules of the puzzle there's no way to do it.

So the second step of the problem is to figure out how to take one ounce from the five-ounce container. One way of doing this would be to get two ounces into the three-ounce one, so that one more ounce would be required to fill it.

That gives us a third question: how can we get just two ounces into the three-ounce container? And that question leads to the solution.

First, he fills the five-ounce container from the jug. Then he fills the three-ounce container from the five-ounce one, which leaves two ounces in the five-ounce measure.

Next, he pours the three ounces back into the jug, and pours the two ounces from the five-ounce container into the three-ounce one.

When he fills the five-ounce container from the jug and then pours enough from it to fill the three-ounce measure, he will have exactly four ounces of vinegar left in the larger container and the three-ounce one can be emptied back into the jug to await the next attempt to solve the puzzle.

Not all number puzzles are so completely logical. If it takes one amoeba exactly two minutes to reproduce itself, and you have two jars of equal size, one containing a single amoeba at the start of the puzzle and the other containing two, how long will it take the single amoeba to divide into enough offspring to fill the jar to capacity if it takes the pair precisely three hours to do so?

This one might look like a difficult exercise in advanced arithmetic, but it's not. Instead, it involves application of the idea of the *geometric progression*.

We know, from the statement of the problem, that the pair of amoebas in the second jar will fill its jar to capacity in exactly three hours.

We also know, from the statement that it takes an amoeba two minutes to reproduce itself, that two minutes after the start of the experiment the first jar will have one pair of amoebas in it—the same as the second jar did at the start.

Therefore, the logical conclusion is that it will take this pair just as long to fill their jar (measured from the time they come into existence) as it will take the other pair to fill theirs (measured from the start of the experiment), so the single amoeba will require three hours and two minutes to fill its jar.

Some number puzzles can be solved by several different techniques, and the fascination of the puzzle then lies in the search for the simplest way of solving it.

For instance, if Mary is now twice as old as Ann was when Mary was the age that Ann is now, and Mary is 24, how old is Ann?

You can convert this one into an algebra problem, and similar puzzles do actually show up in algebra textbooks under the alias of *thought problems* to give students practice in converting real-world situations into systems of simultaneous equations such as those we met near the end of Chapter 4.

If you choose to do it this way, you might use M to stand for Mary's age now, A to represent Ann's present age, and X to indicate the difference in their ages, and write the following series of equations:

1. $M = 24$ (given)
2. $A = 24 - X$ (implied by the fact that Mary is older than Ann.

3. $M = 2*(A - X)$ (given)
4. $M = 2*A - 2*X$ (multiplying out step 3)
5. $M = 2*(M - X) - 2*X$ (substituting for A from step 2)
6. $M = 2*M - 2*X - 2*X$ (multiplying out step 5)
7. $M = 2*M - 4*X$ (combining terms of step 6)
8. $2*M - M = 4*X$ (transposing step 7)
9. $M = 4*X$ (combining terms of step 8)
10. $4*X = 24$ (substituting for M from step 1)
11. $X = 6$ (dividing from step 10)
12. $A = 24 - 6$ (substituting for X from step 11)
13. $A = 18$

This is the technique which an algebra teacher would expect a student to use to figure out that Ann is now 18.

However, there's a more direct method to solve the puzzle. It involves the idea of the *number line* we met earlier. We can draw a line to represent the number of years that have passed, and put a mark on it to show Mary's present age, which is 24. We've done this in Fig. 7-2.

Now, move back toward the zero point of the line and put another mark to show Ann's age at the earlier date, which would be 12. Figure 7-3 shows this step.

Ann's present age is some point between those two marks; and since it is also the point at which Mary's previous age was the same as Ann's present age, it must be exactly halfway between the two because every year added to Mary's age since that time was also added to Ann's age.

Figure 7-4 illustrates this step of the solution. We know that the two points shown in Fig. 7-3 are 24 and 12, respectively, so the point halfway between must be half the difference away from both, which makes it 18.

In fact, this solution to the puzzle makes it clear that all problems of this type involving differences in ages *when A was as old as B is now* must give a solution which is halfway

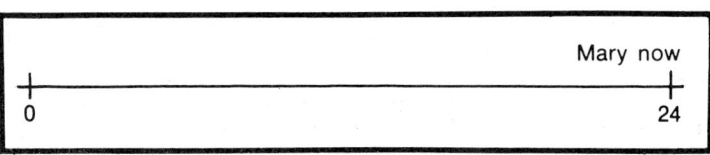

Fig. 7-2. The first step of the number line approach to the age puzzle is to draw the number line and plot in Mary's present age of 24.

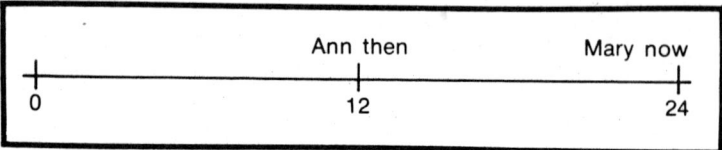

Fig. 7-3. The next step is to add Ann's previous age, which is half of Mary's present age, or 12.

between the present age of the older party, and the multiplying factor (*twice*, in this case).

Not all number puzzles actually involve much use of numbers. Tied up at a dock is a yacht, with a rope ladder dangling over the side into the water. If the tide rises at the rate of 7 inches per hour, and the top eight feet of the ladder are out of the water at low tide, how much of the ladder will be out of the water five hours after the tide begins to rise?

If you try to figure this one out by multiplying the 7-inch-per-hour rise rate of the tide by the 5-hour time span, then subtracting the 35-inch distance from the original 8-foot length, you must have missed the fact that the rope ladder was attached to the boat and rose right along with it—leaving the full 8 feet just as high above the waterline at high tide as it was when the tide was low.

Perhaps the validity of calling this one a *number* puzzle when the numbers are actually there just to introduce a confusion factor is questionable, but it certainly involves some logical analysis of the starting premises.

Our final example of mathematical puzzlement, though, can honestly be called a true puzzle in numbers. If you cover the first half of the distance of a trip at an average speed of 25 miles an hour, how fast do you have to drive during the second half to achieve a total average speed of 50 mph?

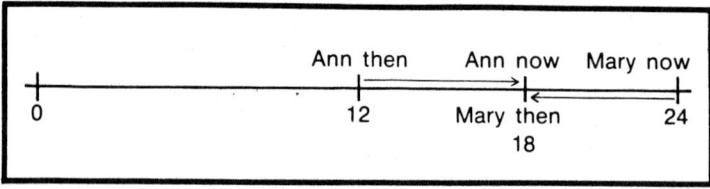

Fig. 7-4. Finally, we plot the point at which Mary was as old as Ann is now. Since every year taken away from Mary's present age to get back to that point must be added to Ann's previous age in order to reach it, this point has to be halfway between the two points of Fig. 7-3, telling us that Ann's present age is 18.

Some folk answer this one by reasoning that the average of any two numbers is the point midway between them, concluding that to raise the overall average by 25 mph to 50 they would have to drive at an average speed of 75 mph over the second half. Since 50 is the average of 25 and 75, their logic appears perfect—but it's wrong.

To see why, pick any distance you like for the trip to be. We'll use 100 miles to keep the numbers simple. This makes the midway point exactly 50 miles. At an average speed of 25 mph, it will take us two hours to drive this 50 miles.

But in order to average 50 mph over the entire trip, we must cover the entire distance in just two hours—and since we're only halfway there at that time, we must cover the remaining 50 miles in no time at all.

This, of course, is not possible, so the only answer to the puzzle is that *it can't be done*.

Chapter 8
Some Branches of Math

Many other fields of interest lie beneath the mathematical umbrella: Some of these are differential and integral calculus, statistics, number theory, more geometry than you could shake a stick at, symbolic logic, topology, and transfinite mathematics.

This volume is far too small to give details of even the few subjects listed above. All we can do in the space available here is try to give you a general idea of these branches of study.

ADVANCED NUMERIC MATH

Mathematicians class such studies as statistics and probability under the generic heading of numeric mathematics. The name comes from the fact that these branches of study deal primarily with manipulation of numbers, or at least produce results which have numeric values.

Nonnumeric areas of study, which we look at a little later, do not necessary produce any kind of numeric results. They depend, rather, on logic and reasoning.

The dividing line between numeric and nonnumeric mathematics, like most dividing lines in this study, is not necessarily as sharply defined as you might think. We have seen in previous chapters how the entire foundation of arithmetic (which, in its turn, is the basis of numeric mathematics) is built upon logic. The grouping, however, serves as well as any other, when classifying the subject into different categories for purposes of discussion.

Something About Statistics

The study of statistics can serve almost any imaginable purpose. Statistical correlations have been used to prove (and to disprove) almost any theory of social studies, public health, or politics, that could be dreamt of. Some of the proofs reached by means of statistics may even have been valid—but the indiscriminate use of statistical proofs is what gave rise to the expression that "figures don't lie, but liars figure."

The collection of statistics dates back to the early 1600s, when tallies of the causes of deaths were originated in the city of London. These tallies made it possible for the authorities to determine the citizens' most pressing problems so far as survival was concerned.

Serious study of statistics, however, was deferred until some time later. During the last decade of the 17th century the astronomer Edmund Halley, who gave his name to Halley's comet, collected a set of statistics covering the previous 5-year period for the city of Breslau, in Silesia, showing the number of births and deaths from various causes, and most important of all, the age at death.

Halley arranged these tables to show the life expectancy for residents of the city, and thus deduced a way in which to determine the probability of a person's survival for the next 12 months.

This stroke of genius gave birth to the life insurance industry.

As insurance became more popular, the collection and study of statistics became ever more essential to its survival. The result was establishment of today's bureaus of vital statistics, and indirectly, the growth of this entire branch of mathematics.

The insurance industry offers many illustrations of the practical application of statistical data.

Let's imagine the probability that a man of a certain age and medical history will die within the next year is determined, from statistical data, to be 1/10. If this person is considering the purchase of life insurance, he should base his decision on that probability—1/10th. The insurance company must, however, consider another probability derived from this same set of data.

If the company is selling policies on a large number of individuals in this same class, the probability is extremely high that the company will have to pay claims on no more than approximately 1/10 of the policies sold. This means that, if the company's rate is rather more than 1/10 of the face value of the policy, it is most likely that enough will be left over after all the claims are paid to meet all expenses and make a profit for the stockholders.

The beauty of this fact from the company's viewpoint is that the greater the number of persons purchasing policies (and consequently, the larger the volume of claims paid) the greater is the probability that the company's finances will stay sound—and the stockholders' profit increase. This is the all-important consideration which distinguishes the insurance business from gambling.

Statisticians deal not with individuals but with populations consisting of large numbers of individual situations or people. The actions or events involving any specific individual are unpredictable, but as more and more are grouped to form a population, the variations which establish individuality cancel, and actions or events involving the mass emerge as clear-cut patterns.

This discovery was Halley's stroke of genius. The specific technique which makes the patterns appear, as if by magic, is that of *averaging*.

What most of us think of when we hear the word *average* is what a statistician knows as the *arithmetic* average or *mean*. That is, it is the value obtained by summing up a number of individual items, then dividing the total by the number of items in the list, to establish the central point about which the individuals vary.

It should be obvious from the nature of the process that within any population there must be as many individuals below average as there are above. Despite this, politicians have been known to entice voters by promising to "bring everybody above average" in living standards.

The idea behind the average is to cancel individual variations, thereby presenting a picture which is true of the majority of the population involved. One nonmathematical application of averaging is the way in which a bowler maintains his or her record of performance.

In league competition, the score bowled by each participant during each game is recorded, and the secretary of the league calculates the bowler's average score after every bowling session. In this case, the population is not that of the bowlers but rather that of the individual scores achieved by one bowler in successive games.

The variations being canceled by the averaging process include those times when the bowler is exceptionally good as well as the nights of miserable scores. The idea is to give, when the average is extended to cover a long enough period, an accurate indication of that bowler's normal performance, and provide a basis for deriving a meaningful handicap in future competition.

Let's work a sample of the kind of calculation involved to get a more accurate idea of the principles of statistical averaging. For instance, if a bowler scores 140, 145, and 150 during the first night of league competition, his total score for the evening will be 435 pins, and his average will be 435/3 or 145.

If, at the next meeting of the league, the same bowler rolls up 181, 190, and 176, his total for that evening will be 547 pins. His average for that evening will be 182.333, but that's not the way it works. His average, instead, is figured from the grand total of pins—the previous week as well as the current week. Thus it is computed from 435 plus 547 or 982 pins, divided by the six games over which they were bowled. His resulting average is 163.666. (In practice, the fractions of bowling averages are dropped. We will keep them in this example because we are concerned with the mathmatical principles involved.)

This individual's performance during the second week raised his standard of performance by $18^2/3$ pins. If, on the third week, he bowled another 547, this average would rise still more.

However, the six-game average tells us enough about the principles involved.

During the first three games, when the bowler established his average of 145, one of the games actually was a 145 score. The other two were each 5 pins off the mark, in opposite directions, and so the "scatter" of five pins canceled itself.

When the total number of games bowled rose to six, the average had become $163^2/3$ and yet not one of the six games contributing to that average was anywhere near that precise score. All three of the first three games were well below the mark, while the seond three games were all well above. The "scatter" of the scores now is 36 pins, from 140 to 176, and the *median* of this scatter is 158—that is, halfway between the high and low individuals of the population.

Note that the median of the scatter is, itself, $5\frac{2}{3}$ pins lower than the average. This comes about because the games above the midpoint were farther *above* it than the submidpoint games were *below* it. The higher-scoring games carried more weight in the average.

The concept of scatter is an essential one in statistics, because it establishes the validity of some of the data involved. If the scatter is extremely large, it indicates that the population of the sample from which the statistics are drawn may be too small to be significant.

In our example, a scatter of 36 pins in an average of $163\frac{2}{3}$ indicates that the sample may be too small to give an accurate picture. If, on the other hand, the scatter had been less than 5 pins from lowest to highest, it would indicate that the average was probably close to that bowler's average performance.

The characteristic of the averaging technique is that it offers a means for eliminating individual variations from a population. The arithmetic average, however, is not the only method of averaging which statisticians have available to use.

Another kind of average, known as the *harmonic mean*, is equal to the reciprocal of the arithmetic average of the reciprocals of the values we wish to average. This is the appropriate technique to use when we want to average rates or prices.

For instance, if we have an airplane flying around a square 100 miles long on a side, and the plane flies the first side at 100 miles per hour, the second side at 200, the third side at 300, and the fourth at 400, what will be the average speed of the plane at its flight (Fig. 8-1) around the square?

If we use the arithmetic average of speeds, we will get $100 + 200 + 300 + 400$, or 1000, divided by 4, or 250 miles per hour as the average.

We can easily show that this answer is wrong. The square is 100 miles on a side. Since the plane flies the first side at 100 miles per hour, it will take one hour to make the trip. The second side, flown at 200 miles per hour, requires 30 minutes. The third side is covered in 20 minutes, at a speed of 300 miles per hour, and the fourth side takes only 15 minutes to traverse at the 400-mile-per-hour velocity.

This gives us a total time of 2 hours and five minutes, to travel a distance of 400 miles, which results in a velocity of 192 miles per hour averaged over the entire distance.

To calculate this average by the *harmonic mean* technique, we would first take the reciprocal of each speed involved. That is, 100 would become 1/100, 200 would become 1/200, and so on. We would then divide four by the sum of these reciprocals.

This would give us the quantity *four divided by 25/1200*, which, when the fractions are cleared out, turns out to be 192 miles per hour, the same as we calculated from our time and distance study.

It is important to note that the *harmonic mean* applies to this problem because the times vary but the distance remains constant. Had the time remained constant, and the distance been variable, the ordinary arithmetic mean would have been correct. The choice of the type of average appropriate to any problem always depends upon the terms of the problem at hand. Caution is mandatory.

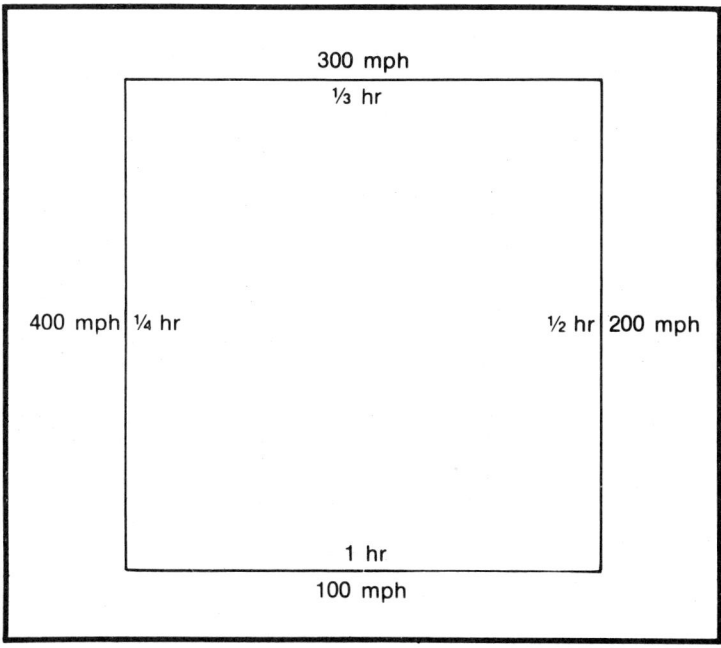

Fig. 8-1. Problem of determining average speed of airplane which goes around this square at different speed for each side is more involved than you might think at first glance.

Still another type of average used by statisticians is the *geometric mean*. This is the type of average which should be used to eliminate individuality from quantities which follow a geometric progression, or exponential law of growth.

The geometric mean is determined by multiplying together the quantities involved, rather than adding them as is done in the arithmetic averaging technique, then taking the nth root if n quantities were multiplied.

That is, where the arithmetic average uses addition, the geometric average used multiplication, and where the arithmetic average used division, the geometric average extracts the corresponding root.

Again, it must be noted that all of these techniques measure a central tendency, which eliminates the effects of individual variations in a population of several individuals.

By eliminating the individual variations, the result tells what may be reasonably expected of a typical member of the population, although it is quite likely that no specific individual within the population exhibits the "typical" characteristics.

Probability and Possibility

Once the statistical lists recording the actual occurences of events had been collected and recorded, it was not an extremely long step to establish the chances that some specific event would occur at some specific time. Man has been a gambler for longer than he has recorded his history.

Although the step was not long, it was fraught with possibility of error and contained a number of subtle logical traps for the unwary.

In Chapter 6 we made some acquaintance with the principles of probability when we examined the odds on the game of chuckaluck. Probability is a ratio, and is typically expressed as a fraction between the limits of *0* and *1*.

A probability of 0 for some event indicates that the event will never occur; a probability of 1 means that the event is certain. Between these extremes, the value of the probability fraction indicates the chance of the event's occurrence.

If there is one chance in two (even odds), such as that which exists in the flipping of a coin, the probability is 1/2 or, expressed as a real number, 0.5000. If the chance is one in three, the probability is 0.3333, and so forth.

The step which connected statistics and probability was Jacob Bernoulli's *law of large numbers*. This states that if the probability of an event occurring is some specific fraction (which we shall call P), then, for a sufficiently large number of attempts, the event will occur in the proportions indicated by P.

That is, if the probability is 0.1, then in ten tries it might occur once—but might easily not occur at all, or occur five of the ten times.

However, if the number of attempts is increased to one million, it is reasonable to expect that one tenth of them—or one hundred thousand tries—will result in the event.

This law of large numbers is the basis of the insurance industry. While no insurance actuary can predict how long an individual policyholder will live, the actuaries know to extreme precision what percentage of their total number of policyholders will become deceased policyholders within the next year—and this permits the rates to be established at a level which the policyholders can afford, yet which assures the company a profit.

One of the traps buried in the laws of probability is the fact that chance has neither conscience nor memory. A gambler who has had a run of bad luck may use the law of large numbers as his excuse for coutinuing to gamble, with the comment that "my luck is bound to change soon."

But probability predictions apply only if the chance, at each attempt, is completely independent of any previous attempts.

That means it can be no more likely for a crapshooter to make his point on the 20th try, than on the second. Any assumption to the contrary violates the basic foundations of probability.

In most games, we can calculate the probabilities involved from the number of possible combinations permitted, just as we did in Chapter 6 for the game of chuckaluck. The probability, for instance, of filling a royal flush (Fig. 8-2) in a poker hand can be computed, since we know that a card deck contains 52 cards, consisting of 13 each in 4 suits, and we know the total number of ways a royal flush can be made—4. Probabilities of this sort are just as certain as are any other mathematical values.

Most of the applications of probability in predictions for games, however, involve another way of determining an

Fig. 8-2. The royal flush is the least likely hand in the game of poker, since only four exist in a 52-card deck (one in each suit), and the total number of 5-card hands possible from that same deck is 2,598,960. This means that only 4/2598960, or 1 in 649740 hands, can be a royal flush. Actual game odds are not quite so extreme, since these figures assume that only one hand is to be dealt from the deck at a time, but the probability of getting one is still extremely small.

event's probability. It is based on statistics, and is based on what is effectively the inverse of Bernoulli's law of large numbers.

For instance, it is not practical to compute the probability of an individual dying at age 25 of bronchial pneumonia. The specific individual involved may never get bronchial pneumonia, in which case the probability of the event is 0, or he may be killed in an auto accident at age 10, which also makes the probability zero that he will die at 25 of pneumonia.

The only way of determining probabilities of this type is to derive the probability value by assuming that the law of large numbers is true, collecting statistics, and from them, computing anticipated probability values.

That is, if you know that in a certain population of 100,000 individuals, 12 died at age 25 of bronchial pneumonia, you can simply use the fraction 12/100,000 (expressed decimally as 0.00012) as the expected probability that any other individual in a similar population will meet his demise in the same way.

When statistics of this sort are kept for insurance purposes, they are called actuarial tables. Actually, the actuarial tables are estimated probabilities (based upon experience) of certain events.

In today's technological world, probabilities play another important role. Almost all of the probability values used in technology are of the kind estimated from statistical data. Great pains are taken to collect the necessary data so that it will produce meaningful results.

When a new airplane is designed, for example, the engineers will build a prototype, and literally test it to destruction in test chambers, accumulating the necessary data to determine the probability of failure of each critical component as a function of elapsed time.

These probabilities, in turn, are used to predict the serviceability of the finished craft and, more important, to establish operating intervals at which components must be replaced.

By replacing a wing support before the probability of its snapping from metal fatigue becomes unacceptably large, the chance of the airplane losing a wing while in flight with a load of passengers is greatly reduced.

This offshoot of the study of probability is known in industry as *reliability*, and is one of the most essential yet least understood facets of the engineering profession.

NONNUMERIC MATHEMATICS

Among the nonnumeric areas of mathematics, we find such studies as geometry, symbolic logic, topology, and transfinite math. All of these share a dependence on logic and reasoning rather than being based on counting, and so are said to be nonnumeric.

Subjects of this sort are usually considered to be higher mathematics, and many folk feel they are far too abstruse for general interest. Nothing could be further from the truth.

Symbolic logic, for instance, brings the discipline and techniques of mathematical methods to bear upon the real-world problem of thinking in a logical manner. While a study of symbolic logic will not make a Mr. Spock out of an Archie Bunker, it can go a long way toward giving anyone a solid basis for his beliefs by assuring that the conclusions are achieved by consistent methods.

Similarly, the study of topology is traditionally based upon its ability to produce existence theorems, which are essential in really high math—but topology is also capable of showing anyone how to construct a sheet of paper which has only one edge and one side. It can even provide us with a magic trick, which we shall see a little later in this chapter.

To put it bluntly, numeric mathematics may have many real-world applications—but nonnumeric mathematics can be plain outright fun!

The Laws of Thought

The publication, in 1854, of George Boole's *The Laws of Thought* has been hailed by hisorians of mathematics as the beginning of pure mathematical thinking. Though many other workers—notably, Augustus DeMorgan and C.L. Dodson (who used the pseudonym Lewis Carroll when he wrote *Alice in Wonderland*)—have contributed mightily to the development of symbolic logic, the Boolean algebra set forth by Boole has formed the foundation of pure mathematics ever since.

We met many of the concepts involved in symbolic logic in Chapter 2, when we learned about sets. The idea of the universe of discourse, the sets and subsets which compose it,

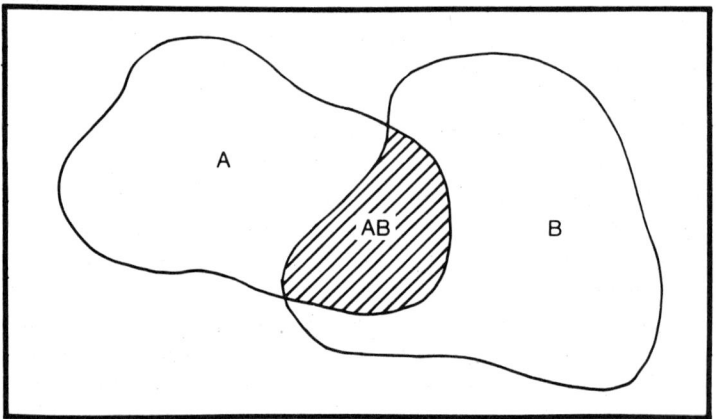

Fig. 8-3. Boole's logical product indicated by expression AB is same as set intersection idea we met in Chapter 2.

their intersections, their unions, and their complements, all were set forth in Boole's early work. Had he done no more than this, he would have laid an imposing foundation for the temple of mathematics. However, he went much further.

The operation which we defined as *intersection* in our look at set theory in Chapter 2 was mapped by Boole into the arithmetic operation of multiplication, and called the *logical product*. That is, the logical product of the classes A and B (Fig. 8-3) consists of their intersection, or the elements which they share in common.

Similarly, the idea of *union* of sets was mapped into the arithmetic operation of addition, and called the *logical sum* (Fig. 8-4).

In both cases, the operator symbol was borrowed from arithmetic. Thus, the logical product of A and B is written symbolically as $A \times B$ or AB, and their logical sum is written as $A + B$. This makes ordinary algebraic manipulation possible.

These operations lead to some startling conclusions. Since the variables represent sets, or classes, rather than numeric values, they do not count up as they would in arithmetic or ordinary algebra.

That is, if A represents all humans over the age of 21, then the logical sum $A + A$ is not the $2A$ which conventional algebra would lead you to expect—because anyone in the class

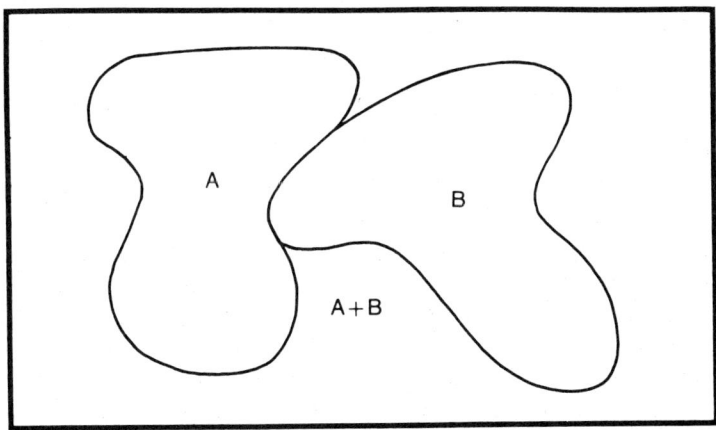

Fig. 8-4. The **logical sum** or **A + B** is the same idea as that of **union of sets**. However, Boolean algebra makes computation of logical implications much easier than using Venn diagrams such as this and Fig. 8-3.

of humans over 21 is already there, and cannot be put there again. If the statement is repeated, it makes no difference.

Thus, in Boolean algebra the logical sum $A + A$ simply is equal to A. The operation can be carried on indefinitely: $A + A + A + A + A + A + A + ... = A$. This is an essential distinction between the techniques of Boolean algebra and those of numeric algebras.

The idea may become a little more clear when you realize that the notation $A + A$ really means *either everyone over 21 or everyone over 21*. That is, the logical sum or *plus* operation in Boolean algebra corresponds to the idea expressed in ordinary language as either/or.

When we introduce a second class, it makes more sense. If we define B as *all residents of New York City*, then the logical sum $A + B$ means *either everyone over 21, or all the residents of New York City*.

This strange behavior of the logical sum might lead you to assume that the logical product would also act in an unexpected manner. Your assumption would be correct. The logical product of A and B is equivalent to the statement A and B, since it is a set intersection. In our example sets, this would mean *everyone over 21 who is also a resident of New York City*.

If we multiply a set by itself logically, so that we have $A \times A \times A \times A \times A$ and so forth indefinitely, the result is still no more than A alone. That is, any class has all its elements in common with itself. In our examples, it means *everyone over 21 who is also over 21*.

Since, in Boolean algebra, no variable need appear more than once in any term of an expression, such things as exponents and coefficients simply do not apply in this calculus. Instead, we have such things as the *law of absorption*, which states that the product $A\ (A + B)$, that is, the collection of elements common both to a class and a larger class including it, is the same as the smaller class, A. This rule helps simplify otherwise unmanageable expressions, and this ability to simplify statements of such relations is the characteristic which gives symbolic logic its unique power to clarify thought.

Let's see how the law of absorption proves out, using the techniques of symbolic logic itself. First, we take the expression $A\ (A + B)$ and multiply it out just as if it were an expression in numeric algebra, to produce the equivalent expression $AA + AB$. Next, we reduce the AA to A, which gives us $A + AB$. When we translate this into words, it comes out as *either the class, A, or that part of the class A which is also in class B*. In other words, either the whole class or some part of it. If any member of the class will do, there is no need to test further—so we reduce the expression to merely A.

Before we go into a more complicated situation showing how easily symbolic logic can clarify a confusing problem, we must introduce a special notation for the complement of a variable.

In most present technical uses of Boolean algebra, the complement of a variable is indicated by overscoring that variable, but this notation is often difficult to apply when setting words into type. For many textbooks and other published material, a *prime* notation is used. That is, an apostrophe-like symbol following the variable indicates that the complement of that variable is meant. Thus, if the class A consists of all persons over 21, then A' or $\bar{A}$ indicates the complement: everyone 21 and under.

This notation is essential to our example problem, which is based on instructions given two librarians who must sort a pile of books just brought back to the library.

The first librarian is told to collect all political works by American authors, and all books by foreign authors longer than 500 pages.

The second is directed to take political works over 500 pages in length and nonpolitical novels written by Americans. Could a conflict arise whereby both try to choose the same book?

Here's how the problem looks when it is set up in symbolic logic:

U = all of the books in the pile (the *universe*)
A = books by Americans
B = books more than 500 pages long
P = political books
N = novels

The complements of the classes are:

$\bar{A}$ = books by foreign authors
$\bar{B}$ = books 500 pages or less in length
$\bar{P}$ = nonpolitical works
$\bar{N}$ = books which are not novels

The first librarian will want all the books which are described by the logical sum of logical products, $AP + \bar{A}B$, or political books by Americans, and books over 500 pages by foreign authors.

The second librarian will want those books described by $PB + AN\bar{P}$, or all books over 500 pages on political subjects, and American novels which are not political.

Any books which fall into the intersection of these two logical sums will be claimed by both librarians. The intersection, or logical product, of the two descriptions is written $(AP + \bar{A}B)(BP + AN\bar{P})$. This expression can be multiplied using the normal techniques of numeric algebra, to give the expression $ABPP + AANP\bar{P} + \bar{A}BBP + A\bar{A}BN\bar{P}$.

The logical product of any class and its complement must be the empty set, since the class and its complement can have no elements in common. This causes two terms ($AANP\bar{P}$ and $A\bar{A}BN\bar{P}$) to vanish from our intersection, since they cannot exist. The law of absorption simplifies the remaining two terms, so that our expression becomes $ABP + \bar{A}BP$.

We can factor this expression, again using normal algebraic methods, to get $BP(A + \bar{A})$. Since the union of any set and its complement is, by definition, the entire universe, and since the law of absorption applies when the larger set is the universe, this expression reduces to simply BP.

The answer to our question is, yes, both librarians will claim books in the class BP, which translates to be *political books more than 500 pages long.*

The instructions given the librarians contain a hidden conflict, and cannot be accurately fulfilled. While the problem could undoubtedly have been solved without using symbolic logic, the use of the symbology and Boole's techniques greatly simplifies both the statement and the solution of this situation.

The Case of the Single-Sided Sheet

At the beginning of this section, we promised a look at one branch of mathematical study which offers the potential for construction of a magic trick. The subject is *topology*, which can be described as *the geometry of distortion.*

Topology deals with those fundamental geometric properties of space which are not changed when we stretch, twist, or otherwise change an object's size or shape.

The name topology comes from the Greek *study of location.* An older name for the same subject is *analysis situ,* which in Latin means *analysis of position*. Topology studies linear figures, surfaces, and solids—its subjects range from pretzels and knots to networks and maps.

Almost all other branches of geometry measure lengths and angles, and therefore earn the description *metric*. Topology, on the other hand, is a nonmetric study—its subjects remain unchanged, whether the objects are made of rubber or are rigid in a plane like the figures encountered in the metric geometries.

Here are some of the problems examined in topology: If a triangle is stretched into a circle, which of its geometric properties are retained? Is the hole inside the donut, or outside? What is a knot? Is it possible to make a bottle that has no edges, no inside, and no outside? Ridiculous as these may appear at first glance, they are examples of serious topological questions.

The study of topology began in the middle of the 19th century. Its orgins, however, go back much further. In 1640 Descartes observed a fundamental relationship between vertexes, edges, and faces of a polyhedron. In 1752, Euler expressed the relationship with the formula $V - E + F = 2$. His publication of the proof of this equation is one of the foundation stones of topology.

Actually, the relationship described by Euler's equation holds true only for shapes which can be related to a ball, a balloon, or other spherical surface. Keep in mind that in topology, we can stretch, squeeze and bend the surface as we like, just so long as we do not cut or tear any part of it. Thus, the Great Pyramid is, so far as topology is concerned, exactly the same shape as a tennis ball (Fig. 8-5).

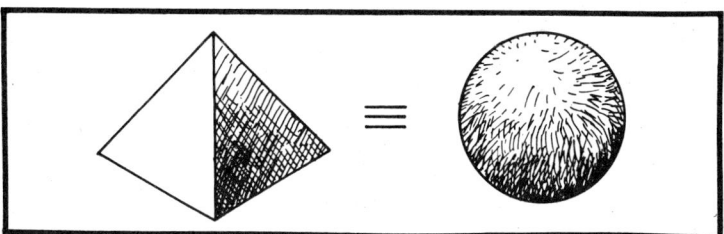

Fig. 8-5. To a student of topology, there's no difference between the Great Pyramid of Egypt (left) and a tennis ball (right), since both are topologically equivalent to a single figure with one outer surface and no holes through it. Difference of size, material, mass, curvature, etc. make no difference in topology.

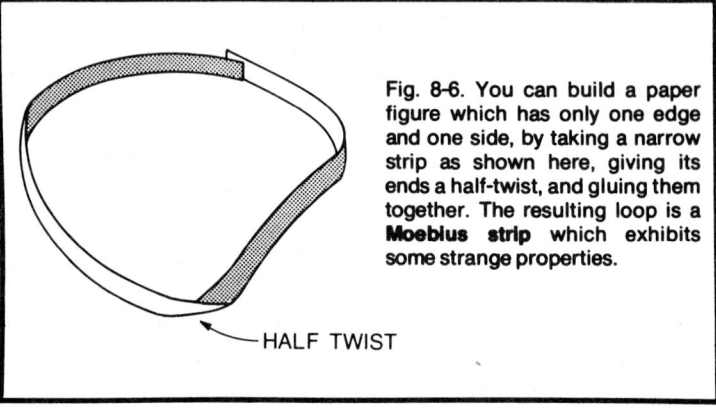

Fig. 8-6. You can build a paper figure which has only one edge and one side, by taking a narrow strip as shown here, giving its ends a half-twist, and gluing them together. The resulting loop is a **Moebius strip** which exhibits some strange properties.

HALF TWIST

The proof of Euler's equation is valid so long as we restrict it to shapes that are topologically equivalent to a sphere. However, if we try to stretch a torus such as an automobile innertube into the shape of the Great Pyramid, we will find that it is not possible, because the innertube shape is not topologically the same as a sphere. The torus has a different relationship. Euler's equation for the torus-shaped figures, is $V + F - E = 0$.

The general form of the relationship, which applies to all possible shapes, is $V + F = E + 2 - 2N$, where N represents the number of holes through the figure. For a sphere, $N = 0$, and we have Euler's equation. For the torus, $N = 1$, and we have the other equation.

Topology is not all a matter of equations. For instance, most ordinary surfaces have two sides. This appears to be an obvious fact, which applies both to closed surfaces like the sphere or the donut, and to surfaces with boundaries such as a disc or a sheet of paper.

We could paint the two separate sides of such a surface in different colors to distinguish them, and if the surface is closed (liked the sphere) the colors would never join. If the surface has boundaries, corners, or edges, the colors would meet only at these edges. For instance, a bug crawling along the surface of a sheet of paper, and prohibited from crossing the edges of the sheet, would always stay on the same side.

The Danish topologist Moebius made the surprising discovery that some surfaces have only one side.

The simplest surface of this sort known bears the name of its discoverer—the Moebius strip, which you can form by taking a long narrow ribbon of paper, giving it a half-twist, then gluing the two free ends together as shown in Fig. 8-6.

A bug crawling along this surface, and dragging a paintbrush with it to mark where it has been, will return to its original position upside down halfway along its trip (Fig. 8-7), and if it keeps going, will eventually end up exactly where it started, yet never crossing an edge. That means this surface has but one side.

You can prove it to yourself by defining side as *that area of a sheet of paper on which you can make a pencil mark without crossing the edge of the sheet*, constructing a Moebius strip, and then tracing a line down the middle of the ribbon. As you trace, you will find that you go all the way around it twice, and both apparent sides are marked when you finish.

In addition to having only one side, the Moebius strip has only one edge. This can be checked, exactly as you checked the side, by marking the edge of the strip. As you trace it, you will find that you pass the point you have already traced, but you are on the "other" edge without ever having left the one you started on.

Now for our magic trick: The ordinary two-sided surface, formed by pasting together the two ends of a ribbon which has not been twisted, has two distinct sides and two distinct edges. If you cut down the middle of such a pasted ribbon with a pair of scissors, it will part as two separate loops.

If a Moebius strip is cut in exactly the same way, it will remain in one piece. The magic trick consists simply of

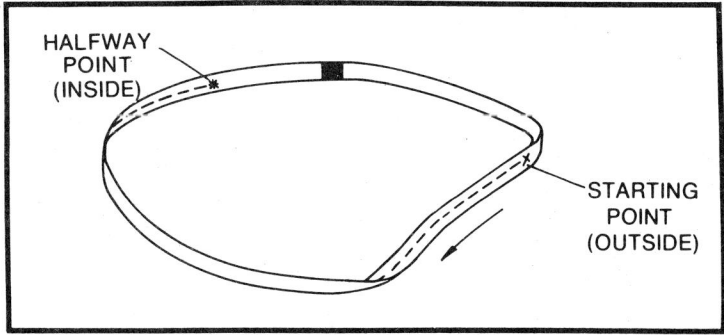

Fig. 8-7. If a bug were to drag a wet paintbrush around a Moebius strip, he would find that he was on the "other" side of it after half his journey, although he never left the side on which he started his trip. This is the proof that the strip does, indeed, have only one side—even though your eyes can clearly see two!

preparing two apparently identical loops, one of which is a Moebius strip, and the other of which is not. First, you cut the non-Moebius loop, demonstrating that such a loop, when trimmed or sliced, falls into two distinct pieces.

You then invoke the magic word *Moebius*, cut the other strip—and, as if by magic, it remains a single piece after being cut.

This strange behavior of the Moebius strip is so opposed to what all knowledge and experience predict should occur, that most observers who are not already aware of the fact refuse to believe that it is possible. But we're not through yet. If the single strip that results after the Moebius strip is sliced down its middle is *again* sliced down the middle, two separate loops will result—linked together like the links of a chain.

This study of topology involves many more facets than we have hinted at here, but like the other advanced subjects, they require more detail than we can afford in this space. Newman's *The World of Mathematics*, referred to elsewhere in this volume, contains an excellent introduction to the subject.

The Arithmetic of Infinity

If your mind was boggled by the idea of a piece of paper having only one side and one edge, get set for another shock. One of the most far-out branches of nonnumeric mathematics is the one called *transfinite math*, which deals with the arithmetic of infinities.

Note well that the word is *infinities* and not "infinity." While in everyday numeric mathematics, as we emphasized earlier, the idea of infinity is that of a region beyond all boundaries, or a number larger than anything can be. In transfinite math it is a *class* of such places rather than just *one* place.

Transfinite mathematics, more so than most branches of the language, illustrates the difference between the *ideal universe* of mathematics and the realities of the universe in which we live and work. It deals with ideas which can never be experienced and which—because they are so strange —are rejected by many students.

The foundation of transfinite math is simple enough. We went through it in Chapter 3 when we found out how numbers are defined by the common characteristic of several sets which are equivalent. Thus, the number 2 was defined as that characteristic common to all sets with more than one element and less than three (Fig. 8-8).

All other natural numbers are defined in the same way. They are known as cardinal numbers when so defined. The step from the everyday set of natural numbers N into transfinite mathematics is the assertion that in the definition of cardinal number, the finiteness of the sets considered is in no way involved. That is, transfinite math claims that the definition of cardinal number can be applied as readily to infinite sets as to *finite* sets.

When this assertion is accepted, the concepts of *equivalence* and *cardinal number* thereby transfer to sets containing infinitely many elements. No such set can exist in

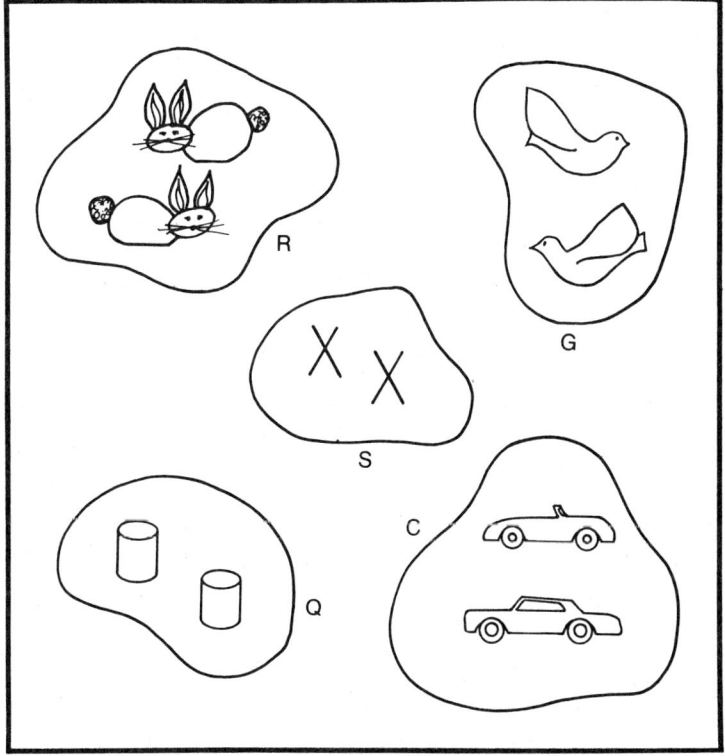

Fig. 8-8. All these different sets—the pair of rabbits R; the pair of gulls G; the two X symbols, S; and so on—are equivalent so far as their number of members is concerned. This equivalence defines the concept of the cardinal number 2. Similar reasoning is used to define transfinite cardinal numbers.

reality, since any existing set must have a limited number of elements in order to draw the boundary around the set which makes it a set.

However, the imagination is not as limited as reality, and we have dealt comfortably with infinite sets for some time now. For instance, the set of all natural numbers consists of the numbers 1, 2, 3, 4, 5, and so forth, but it has no upper limit. No matter how large a number you think of, you can always add one more to it, and have a new largest number.

Similarly, the set of rational numbers is unlimited at its upper end, and so are all the other number sets. It is not just the number of sets which are unlimited. The sets of all the geometric points on a line, lines on a plane, and planes in a space, are equally unlimited.

You can see that the idea of infinite sets has been with us for some time, although intuitively it may appear to be a contradiction. The contradiction is made apparent if we recall that infinity can also be translated as *beyond limits*, and the act of establishing or defining a set is the act of drawing limits around it. This makes the phrase *infinite set* equivalent to the phrase *unlimited limits*.

However, this offers no particular problem to transfinite mathematics, because the ideas with which it deals are (at least within themselves) logically consistent.

We established the cardinal numbers by placing sets in equivalence to each other, and then determining the characteristic which makes them equivalent. Thus, all single membered sets have the quantity of members 1, which defines the cardinal number 1. We carry this out until we have established the rule for generation of cardinal numbers, and that rule defines the set of natural numbers.

We have just defined our first infinite set—the set of all natural numbers, or n. If any other sets exist which are equivalent to this set (in the same sense that all five-membered sets are equivalent and thus establish the natural number 5), the cardinal number of elements in all these equivalent infinite sets will establish the cardinal number of at least one infinity.

That is, if we can establish one-to-one correspondence between the set of natural numbers and any other set, the other set must then—by the definition of one-to-one correspondence—have the same number of elements as there are natural numbers. This leads to some interesting developments.

The act of placing elements of a set in one-to-one correspondence with the set of natural numbers is usually called counting, since the act of counting objects is an enumeration of the numbers involved.

That is, if we count the number of grains of sand on a beach, we assign the number 1 to the first grain we pick up, the number 2 to the second, the number 3 to the third, and so on until we run out of either beach or patience.

Thus, any set whose elements can be counted—or, begun to be counted, since we could never reach the limit of an unlimited number—can be said to be *denumerably infinite*. This means it has the same number of elements as the set of natural numbers.

One denumerably infinite set of this sort is the set of all even natural numbers. At first glance, you would expect that there could be only half as many even numbers as there are numbers in all, since every other number in the set of numbers is odd. This however, is not the case.

Since we can speak of the first even number, the second even number, the third even number, and so forth, we can establish one-to-one correspondence (Fig. 8-9) between the set of even numbers and the set of all natural numbers, so the cardinal number of the two sets is identical.

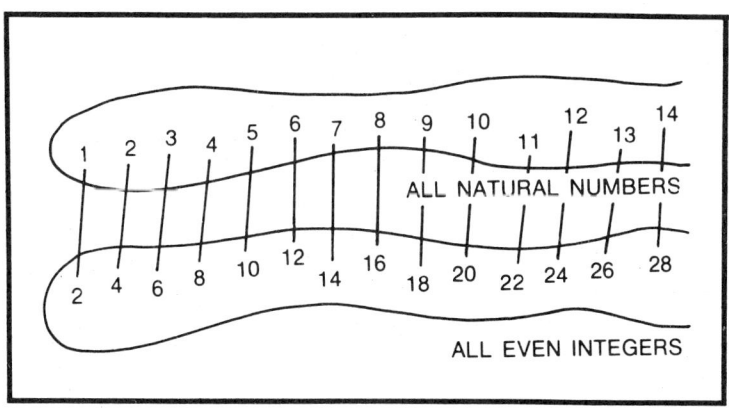

Fig. 8-9. Two infinite sets which can be matched up, element for element, so far as you care to do so, such as the set of all natural numbers at top and that of all even integers at bottom, are said to have the same (transfinite) number of elements. This reasoning is the cornerstone of transfinite math.

205

The same, of course, is true for the odd numbers. We have now established that in the case of even and odd numbers, the part is equal to the whole, and the sum of the parts is no greater than either part alone. This appears to contradict all the common-sense rules of arithmetic—but actually, it merely underscores that the arithmetic of infinity is not ordinary arithmetic.

One of the leading mathematicians of the 20th century was the German David Hilbert, and he illustrated this paraxodical property of infinite arithmetic with the following story:

Let's imagine a hotel which has a finite number of rooms, all of which are occupied. A new guest arrives and wants a room. Sorry, says the manager, but all the rooms are occupied.

Now let's imagine a hotel with an infinite number of rooms; again, all of them are occupied. To this hotel, also, comes a new guest who wants a room.

"But of course!" exclaims the hotel keeper. He moves the person previously in room 1 into room 2, the person who was in room 2 into room 3, the one who was in room 3, into room 4, and so forth. After all the moves are finished, room 1 is vacant, and the new customer gets it.

That takes care of a single new arrival, but let's imagine that now an infinite number of new guests come in and ask for rooms. "Of course," says the manager. "Just give me a little time."

He then moves the occupant of room 1 into room 2, the occupant of room 2 into room 4, the occupant of room 3 into room 6, and so forth. This leaves all the odd-numbered rooms vacant, making way for the infinity of new guests.

With this example, Hilbert drove home the point that operations on infinite numbers involve properties rather different from those to which we are ordinarily accustomed.

When we apply these rules further, we discover that the number of all rational numbers is exactly the same as the number of all natural numbers. We do it exactly the same way. That is, we establish a one-to-one correspondence between the unlimited number of *fractions* and the unlimited number of *integers*. Thus, the cardinal number of these two infinite sets is the same.

So far, all that we have proven logically is that the cardinal number of all the infinite sets we have examined is the same. Could this perchance mean that all infinities have the same cardinal number?

Fortunately for the study of transfinite mathematics, no. At least two higher orders of infinity can be demonstrated.

One of them is the number of all points on a line. The illustrations which prove that no equivalence can be set up between the points on a line and the rational numbers is the argument which establishes the existence of the real number set, in Chapter 3.

No matter how many points are occupied by rational numbers, there are still points left over (corresponding to the reals). This means that the set of real numbers, the set of points, and any other sets which may be equivalent to them, represent a higher order of infinity than the set of natural numbers.

Transfinite mathematics was established by Georg Cantor in the years 1871–1874. Cantor first developed the theory of sets, and moved from there into the study of the properties beyond infinity.

He named the infinite cardinal numbers *aleph*, from the Hebrew word for *one*. The cardinal number of the set of natural numbers, the lowest order of infinity we have examined, he called *aleph-ull*. The cardinal number of the set of all points on a line, or the set of all real numbers was designated *aleph-one*.

Students of transfinite mathematics have shown that aleph-one is the number of all geometric points on a line, in a square, or in a cube. And Cantor proved that it was equivalent to the number of all points in space, no matter how many dimensions were involved.

It might seem that no higher order of infinity could exist—but it has been found that the variety of all possible geometrical curves, including those of most unusual shape, has a larger membership than all the geometrical points, and therefore requires the number *aleph-two*. No other example requiring *aleph-two* has yet been announced.

Now that we have learned what is meant by the *arithmetic of infinity*, a couple of questions arise. Does this arithmetic exist in mathematics? Does it have any use?

To answer the first, it certainly exists within the language of mathematics, or else we could not have explained it to the degree we did in the preceding paragraphs.

While certain contradictions are implicit in some discussions of the so called infinite sets, that does not indicate that set theory or transfinite mathematics is faulty. It simply indicates that certain expressions which can be expressed in the language of set theory *should not be*. The situation is exactly equivalent to that introduced in everyday arithmetic by attempts to divide by zero, and the solution should be the same—to outlaw the forbidden expression and label the result meaningless.

The type of contradiction involved is illustrated by the statement: *This statement is false*. If the statement is true, it must be false, and vice versa. Since the statement is impossible to evaluate, it must be considered meaningless.

Expressed in the language of set theory, a similar statement involves *the set of all sets which are not members of themselves*. The question is, *Does it, or does it not, belong to itself?*

If it does, it cannot, and vice versa. Like the unevaluable statement of truth or falsity, it must be considered a meaningless expression.

No firm answer to the second question (*Is it of any use?*) can be made—because transfinite mathematics is, today, in the same state of development that symbolic logic enjoyed a century ago.

In the last century, the purity of symbolic logic has become sullied by everyday applications. Today transfinite mathematics is essentially pure (that is, incapable of being applied to the real world), but who knows what the next century may bring?

Chapter 9
A Whirlwind
Look at Calculus

Most people tend to think of calculus as one of the most complicated subjects mathematics has to offer. Actually, all mathematics involves some sort of calculus, as we shall soon see. The word *calculus* means, in mathematics, method, and comes from an ancient Greek word for pebble or rock.

For this reason, almost every branch of mathematics is technically a calculus. Most nonmathematicians, though, when they speak of calculus, have a specific mathematical method in mind, and in this chapter we take a look at it, and at several other stones.

CALCULI OF CALCULATION

As we just noted, the Greeks had a word for it—calculus, meaning a smooth stone or pebble, like those you may find on the shores of the Aegean Sea, where formal mathematics as we know it originated.

The ocean-polished pebbles were used as tally markers by some of the mathematicians, and from this use, all mathematical techniques came to be known as *calculi*, the plural of calculus.

The Calculus of Arithmetic

The phrase calculus of arithmetic may strike you as a bit presumptuous, or even downright odd. However, ordinary everyday arithmetic *is* a mathematical technique, and mathematicians know it technically as a calculus. Any technique used for calculation is a calculus, and arithmetic is one such technique.

In Chapter 5, we examined in some depth the rules for arithmetic, and developed them without mentioning calculus, but what we explained at that point was the calculus of arithmetic. The big words are not normally used by elementary school teachers, because they don't want to frighten their students.

The Calculus of Algebra

The calculus we know as *algebra* is similar to, but different from, the calculus of arithmetic. The significant difference is that in algebra, variables enter directly into the mathematical statements rather than being confined to the left-hand side of the equation which states the problem. In simple arithmetic, only constants are acceptable as operands within the statements of the problem.

This difference makes the calculus of algebra different from that of arithmetic. Specifically, the difference between algebra and arithmetic is that each arithmetic statement is a special case, and an algebraic statement is more general.

For example, the arithmetic statement $1 + 2 = 3$ is true, and defines the value of the constant 3; but it tells us nothing about the values produced by adding any value other than the constant 2 to 1. The similar algebraic statement $1 + x = y$ assigns no definite value at all to x or y, but does tell us that whenever we know a value for x, the value of y is one unit greater. In this sense, the statement is true for all values of x rather than just a single value, and so has greater generality.

Mathematicians often compare two or more methods in terms of their relative power. The concept of mathematical power is not strictly defined, but one measure of power is the generality involved.

If technique A applies to a million situations, while technique B applies to only 10, A is said to be more powerful than B. In this sense, algebra has more power than arithmetic, although arithmetic is still necessary, once the algebraic calculation is completed, to assign definite values to the result.

The name *algebra* comes to us from the time and place of the *Arabian Nights*. Caliph Harun-Al-Rashid, (the caliph of the

Arabian Nights), had the mathematical classics translated into Arabic, and encouraged mathematical activity in general.

His son Al-Manun, who succeeded him, was not only interested in mathematics as a science, but was in his own right an accomplished astronomer.

The greatest of Al-Manun's associates was Al-Khowarazmi. Al-Khowarazmi wrote a mathematics book using Hindu numerals, which introduced this set of numerals to Europe and gave them the name by which they are known today—Arabic numerals.

Another of Al-Khowarazmi's writings was titled *Transposition and Removal of Terms in an Equation* (in Arabic, *Ilm Al-Jabr Wall Muquabalah*). People referred to this work by the second word of its Arabic title, *Al-Jabr*, which, when pronounced, comes out *algebra*.

ANOTHER LEVEL OF ABSTRACTION

Many mathematicians consider their subject to be not a science but a form of art. They speak of elegance and beauty of mathematical theorems, proofs, and equations, and judge their results more by questions of following the rules than by standards of applicability to the real world.

For instance, Einstein's special theory of relativity originated as an exercise in mathematical art, and subsequently Doctor Einstein discovered that the form of the exercise would account for an anomaly in practical physics. The result was the *Special Theory of Relativity*, which turned physics upside down (or right side up, depending upon your frame of reference).

This is an almost ideal example of development of mathematics for its own sake, followed by discovery of an application in the real world. What is not obvious, from this example, is the fact that *most* mathematical progress follows this schedule.

Let's look at this question of art in mathematics from several angles, and see how it applies.

Graphic Art and Algebra

Back in Chapter 3, when we looked at the various kinds of sets of numbers available, we made the acquaintance of the *number plane*, which provides a separate point for every conceivable complex number.

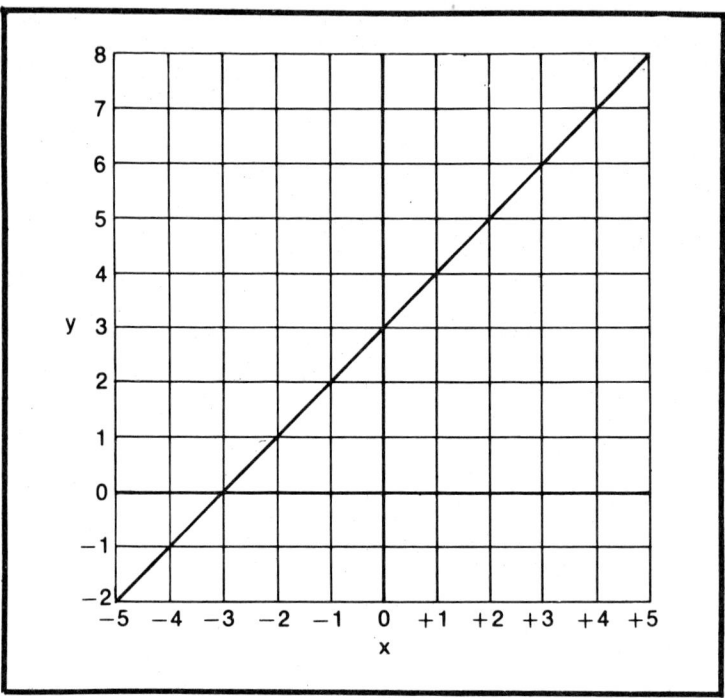

Fig. 9-1. Equation **Y** = **X** + **3** produces graph which is straight line. Values of **Y** are shown on vertical axis; those of **X** on horizontal. Angled line is location of all points for which the equation is true.

Instead of using the operand i for the vertical axis of the plane's coordinates, we simply use two intersecting number lines (we usually call them x and y, with y in the vertical direction and x in the horizontal). We then have a surface upon which we may graph any algebraic equation involving two variables.

For example, we can take the equation $y = x + 3$. If x is 0, y is 0+3 or 3. If x is 1, y is 1+3 or 4. If x is 2, y is 2+3 or 5. The value of y depends entirely upon the value of x, and for every value of x, a corresponding value of y exists. Figure 9-1 shows the graph of this equation for values of x in the range from -5 to $+5$.

Note that the line representing all the values of y is straight, at a constant angle with both coordinate axes, and crosses the x axis and y axis at only one point each. These points are called the *intercepts* of the equation—the y intercept is the point at which the line crosses the y axis; at this point, x equals 0. The value of y in this case is 3.

The x intercept, similarly, is the point at which the line crosses the x axis. At the x intercept y is 0, and the value of x is -3.

Since the graph of this equation is a straight line, the equation is termed *linear*. Not all linear equations involve the same simple addition scheme. For instance, the equation $y = 2x$ is linear, as shown in Fig. 9-2. In this case, both the x and y intercepts are at the origin. When either variable is 0, the other must be also, since anything multiplied by 0 is 0. The equation is linear, but has a different angle with respect to the coordinate axes than the simple additive case.

When we introduce powers into an equation, its graph is no longer linear. The graph of the equation $y = x^2$ appears as Fig. 9-3. If $x = 0$, $y = 0$. If $x =$ either $+1$ or -1, y equals 1. For values of $x = +2$ or -2, the value of y is 4, and so forth.

The specific curve generated by this equation is called a *parabola*, and finds wide application in practical physics. Most automobile headlight reflectors, for instance, have a parabolic

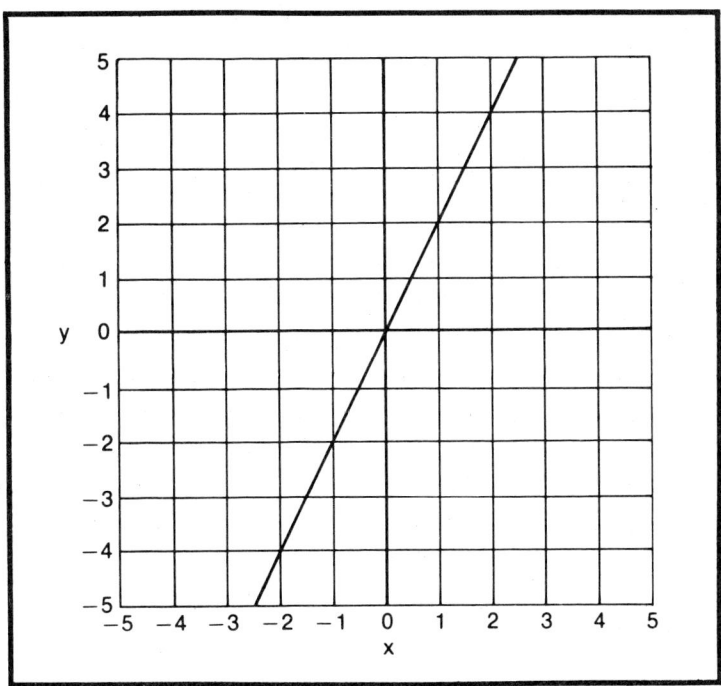

Fig. 9-2. Straight line also results for equation **Y=2X**, but angle between line and axes of graph is different. Note also that this line passes through point 0, 0, or origin of graph; that of Fig. 7-1 did not.

213

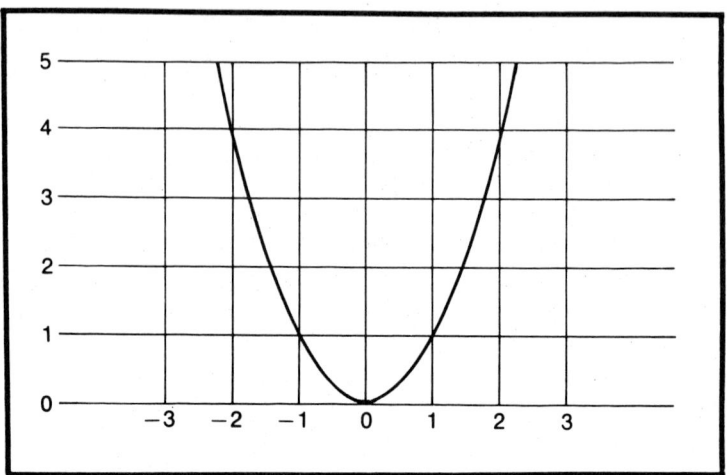

Fig. 9-3. Not all equations graph into straight lines. The curve shown here is the graph of the equation $Y = X^2$, for values of Y smaller than 5. The graph of the full equation is infinitely tall, since values for X and Y are not limited.

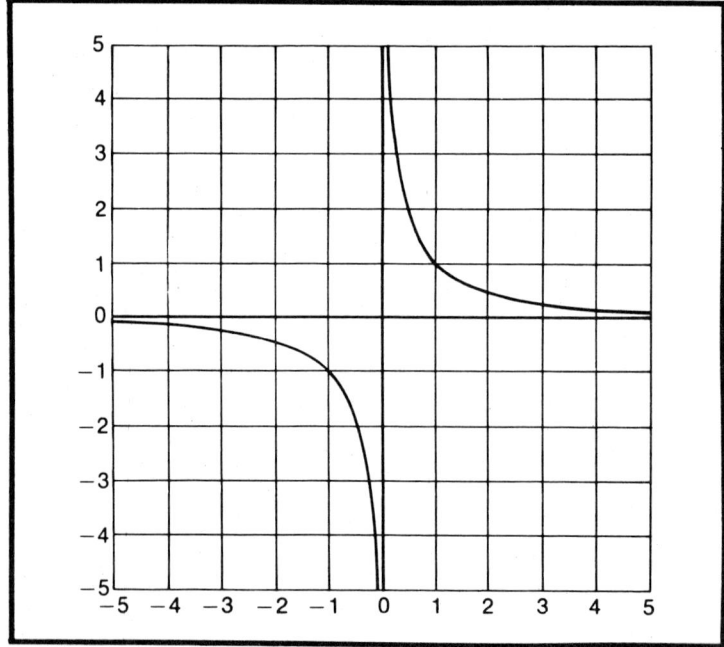

Fig. 9-4. Reciprocal function $Y = 1X$ produces rectangular hyperbola shown here. On large enough graph, the curve looks like a pair of right angles, whence comes the name rectangular. Like the parabola of Fig. 9-3, the hyperbola is infinite.

curve as their cross section, to direct the beam of the bulb into a bundle of parallel beams.

Other equations have different curves. For instance, the equation $y = 1/x$ produces a curve known as the *rectangular hyperbola*, which is shown in Fig. 9-4. Both the hyperbola and the parabola are infinite or unlimited curves. Both have open ends; in the parabola, as x increases without limit so does y (in the hyperbola, y decreases as x increases, but still without limit).

To get a closed curve such as a circle as shown in Fig. 9-5, we must introduce more than one operation into our equation. For instance, the equation for the circle is $x^2 + y^2 = $ (constant).

The graphs we have seen so far involve relatively simple equations. Not all graphs are so simple. For instance, Fig. 9-6 shows a curve known as the *Folium of Descartes*. The equation of this curve is $x^3 + y^3 = 3xy$. As the equations involve functions of ever more complex nature, the curves too become more complex.

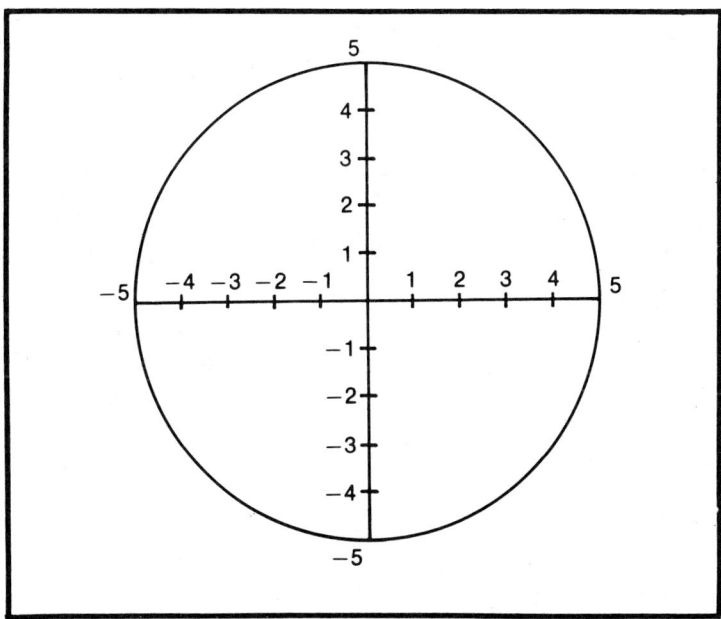

Fig. 9-5. Even the circle has its equation, and so can be graphed. In plane geometry, the circle is defined as the locus of points equidistant from a common center. The equation is $X^2 + Y^2 =$ **constant**; the circle shown is the graph of the equation $X^2 + Y^2 = 5$.

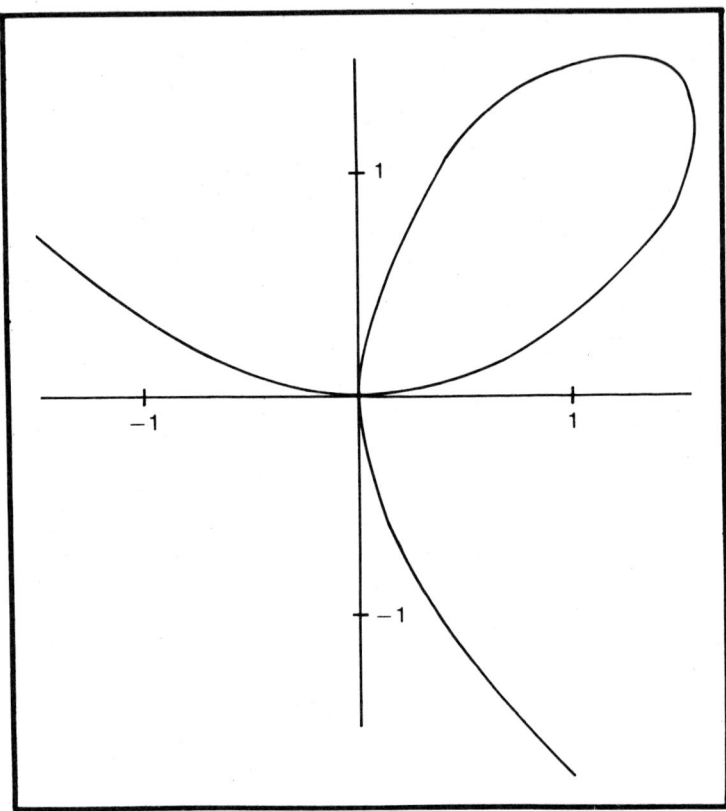

Fig. 9-6. Complicated equations can produce complicated curves when they are graphed. The folium of Descartes is produced by equation $X^3 - 3XY + Y^3 = 0$, and produces this loop for values of **X** and **Y** close to zero. The graph shows the range -1 to $+1$ for both variables.

The study of such material is called *analytic geometry*, and represents a blending of the ideas of algebra with those of geometry. In itself, it is useful to engineers and physicists; to students of mathematics, the primary usefulness of analytics is an introduction to the ideas involved in higher-order calculi.

Limits and Logic

When we examined the method for numbering infinity, in Chapter 5, we became acquainted with mathematical series. We found that two kinds of series exist: one converges to some limit, while the other, which has no limit, diverges.

For example, the series, $y = 1/1 + 1/2 + 1/4 + 1/8 + 1/16 + ...1/2^n...$ converges to a limiting value of 2. If we take only the first term, the value of y would be 1. The first two terms give a value of 1½; the new value is halfway between the old value and the limit.

With three terms, the value is 1¾; with four, 1⅞. Each additional term takes us halfway through the distance left to go. No matter how many terms we take, though, we never reach the limit.

With 15 terms the value of y is $1/2^{16}$ smaller than 2. With 100 terms, the difference between the value of y and 2 is $1/2^{101}$. To reach the limit requires an unlimited number of terms, but the limit exists.

On the other hand, the series, $y = 1 + 2 + 3 + 4 + ...N + (N + 1)$ ...diverges. Every additional term included increases the sum by a larger amount than did the previous one; the sum increases with no upward bound.

A series may diverge in another way. The series $y = 1 + 1 - 1 + 1 - 1 + 1 - 1...$ has the value of 1 for an odd number of terms, and 2 if the number of terms is even. In this case, no limit exists and the series diverges.

Some functions diverge over part of their range, and converge elsewhere. For example, the function $y = 2 + 2x + 2x^2 + 2x^3 + 2x^4 + $ ...$2x^N$ becomes, when $x = 1$: $y + 2 + 2, +2, +2, +2 ...$; this series diverges. If $x = 1/2$, the same function becomes $y = 2 + 2/2 + 2/4 + 2/8 + 2/16 + ...$, which converges to the limit 4. Thus, for values of x less than 1, the function converges; if x is equal to or greater than 1, the series has no finite limit.

The idea of limit which we encountered in that brief look is one of the more abstract, yet essential, concepts of higher mathematics. Just to show how important it is, think back to the problem of fencing the pasture, which we worked out in Chapter 6. There we applied the limit idea to geometrical shapes, and found as a result that the circle encompasses more area than any other figure having the same circumference. We reached this conclusion by starting with a square, shaving off corners to get an octagon, and repeating the corner shaving until we reached the limit of having no more corners to shave.

In Chapter 5, just before we made the acquaintance of the limit idea, we met functions in general. There, we found that any function can, in concept, be considered as a *table lookup* operation, although many practical functions cannot be accurately represented in a finite table.

Now it's time to combine the idea of the function with the idea of limits, to develop a mathematical definition of the limit idea. A function is said to approach a limit, as its variable approaches some value, if the value of the function, minus the limit, becomes and remains less than some preassigned quantity.

Ideally, the quantity preassigned should be zero; but this would imply absolute accuracy. In practice, the preassigned value which establishes the limit is some value so small as to be considered insignificant.

A mathematician, speaking of the limit concept, would say the function of x approaches b as x approaches a, and might then say the limiting value of the function of x as x approaches a, equals b. He would write this as "$\lim f(x), x \to a, = b$."

Three facts concerning limits are of critical importance in our progress toward the higher realms of mathematics. The first is that the limit of any finite sum is equal to the sum of its limits. That means, if the limit of function f is equal to b, and that of function g is equal to c, then the limit of the sum of functions f and g is equal to the sum of the limits of b and c.

Similarly, the limit of a product is equal to the product of the limits.

The third major point is that the limit of a quotient is equal to the quotient of the limit, *provided* that the limit of the denominator is not zero. If the limit of the denominator equals 0, the limit of the quotient is undefined.

This idea of the limit of a function is used to determine whether a function is continuous at any given point. For example, the function $y = x^2$ is continuous at the point $x = 0$, since the limit of the function as x approaches zero is equal to the value of the function at zero.

If the limiting value is *not* equal to the actual value at the point, the function is said to be *discontinuous* at that point.

Many functions are continuous over *part* of their range and discontinuous at other points *within the range*.

For example, the function $y = 1/(x - 1)$ exhibits strange behavior. For a value of $x = -999$, the value of y is $-1/1000$. As x increases (that is, becomes more positive, or moves to the right along the number line) the value of y goes more negative. When x reaches a value of 0, y is -1. When x is $+1/2$, y is -2.

As x approaches the value 1, the value of y becomes increasingly more negative. When x is equal to 1, y is equal to *1 divided by 0*—an incalculable quantity. If, however, x continues to increase, the value of y can again be calculated. Now, it is a very large positive quantity, which decreases as x grows larger. Figure 9-7 shows this relationship between the values of x and y.

The function $y = 1/(x - 1)$, is discontinuous at the point $x = 1$, but continuous at all other points in its range. It is only one of many such functions.

Possibly the most famous such discontinuity is the one in Einstein's theory of relativity, which declares that the speed of light can never be exceeded because mass and inertia become infinitely large as velocity approaches that limit.

A function is said to be continuous in any interval (range between two specified points) if it is continuous at every point within the interval. It may be discontinuous at either or both of the points marking the ends of the interval. Mathematicians call this an *open* interval.

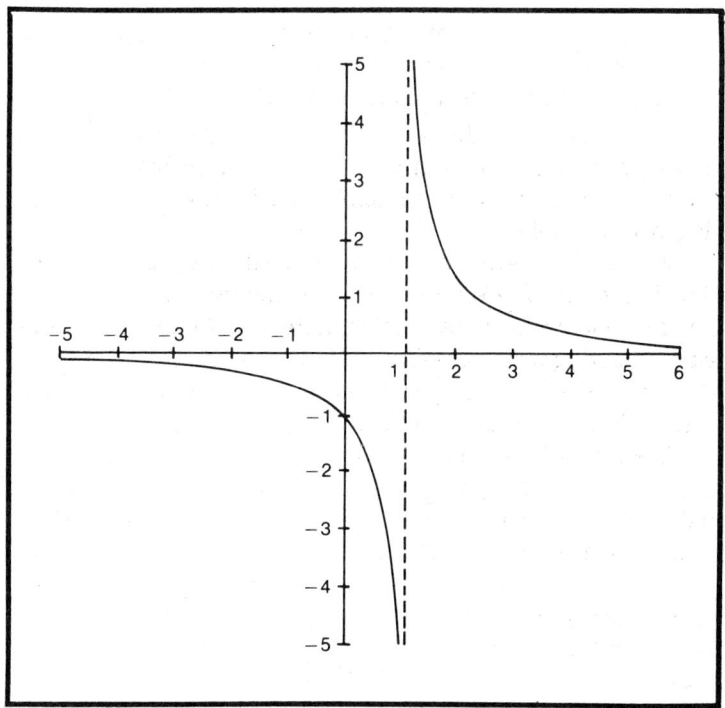

Fig. 9-7. Function **Y=1/(X − 1)** has discontinuity at point **X=1**, but is continuous over the rest of its range, as shown by this graph. In immediate neighborhood of discontinuity, value of **Y** is nearly infinite.

If we have become sufficiently confused with this discussion of the logic of limits, maybe it's time for us to back off from the rarefied atmosphere of mathematical language and look at the whole subject from a more practical viewpoint. Don't worry—we'll put it all back together before very long.

Rate of Change

Let's imagine that you are going for an automobile trip. You leave your home, drive through traffic for a while, and the traffic is really heavy. Sometimes you are going 15 miles per hour, sometimes you are stopped for red lights, and sometimes you don't go as fast as 15 miles per hour, with the result that you are only five miles away from home when a half-hour has gone by.

You fight it out for another 20 minutes, sometimes doing 20, sometimes 10, and then you decide to give up and let the traffic thin out. You stop at a roadside restaurant and have a cup of coffee.

By now, you are 10 miles from home, and it has taken you nearly an hour to get there. You spend 15 minutes drinking your coffee, and the traffic clears somewhat while you do.

When you start again, you're able to speed up from 15 to 35 and eventually reach 60 miles per hour on an open country road. Two hours away from home, you have covered a total distance of 30 miles.

Figure 9-8 shows a graph of your progress on this imaginary trip. The solid line is an indication of exactly how far from home you are at every minute of those two hours that elapsed during our illustration. The numbers scattered along that line represent the speedometer reading you might have seen at that particular point along the way.

Since it took you two hours to get 30 miles away, your average speed was 15 miles an hour. The straight dotted line indicates this average speed. You can see that for small segments the solid line is parallel to the dotted line, but most of the time it is at a different angle to the axis. Part of the time—when you are stopped, having coffee, or at rest at the traffic lights—the actual speed is zero (represented by a horizontal line).

This graph illustrates the difference between *average* and *instantaneous* speed as indicated by your automobile speedometer. If you have ever made a long trip by automobile, you know that the average speed is always much less than you would expect from the speedometer readings, since the times when you are not moving detract seriously from the overall average.

The speeds indicated by your speedometer, charted as the solid line in the figure, indicate the *rate of change of distance with respect to time*, which is what the speedometer indicates. The greater your velocity, the greater this rate of change. In the graph, a higher speed is indicated by a line more nearly approaching the vertical. Thus, in the last five minutes of the graph, when the speed approaches 60 miles per hour, the line is at its steepest angle.

This idea of rate of change is essential in many practical applications of mathematics, especially in the field of physics. The physical quantities involved in velocity and acceleration are simply distance and time. These are the two quantities represented on the two axes of the graph in Fig. 9-8.

Any motion away from the origin changes the distance involved; time changes naturally. Thus, any moving object will have some rate of change of distance with respect to time, or velocity. We usually call this its speed. All that the velocity

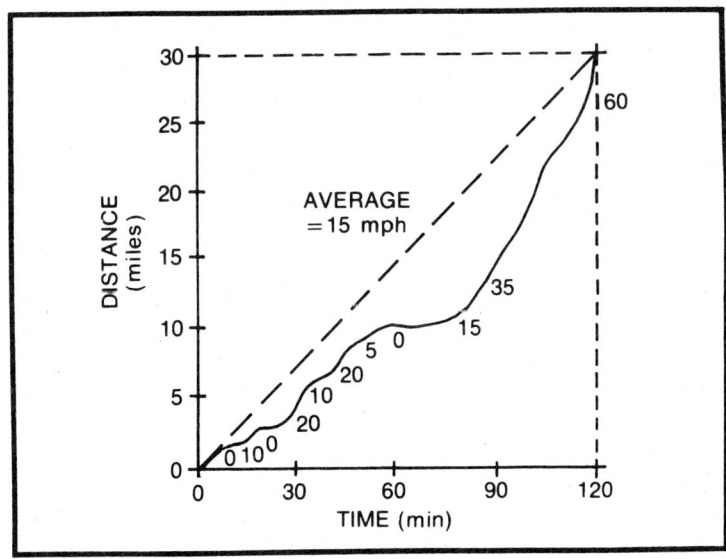

Fig. 9-8. This graph is not related to an equation. It charts **elapsed time** against **distance covered** during an imaginary auto trip. Dotted line represents average speed; figures on solid line show actual speed at various times.

is measuring, however, is the rate at which the distance is changing as time passes.

When we change the velocity, it also changes at some definite rate. This brings us to the idea of the rate of change *of a rate of change*. The rate of change of distance with respect to time, we call *velocity*. The rate of change of velocity with respect to time, we call *acceleration*.

To draw the graph shown in Fig. 9-8, we could simply record the distance measurements (indicated on the odometer of an automobile) at regular time intervals, and that would give us the desired information.

While driving we use a mechanical analog computer called a *speedometer* to perform this mathematical function for us, and indicate our velocity at all times in order to comply with the speed laws.

To determine from the graph alone what the velocity is at any point, it is necessary to determine the rate of change at that point. To do this, we follow the same basic procedure used to determine average speed over the two-hour interval. However, we shorten the time interval.

Rather than averaging over a two-hour period, we shorten the interval until it becomes so short that the average undergoes no change. That is to say, we determine the limit of the average (or velocity), as the elapsed time approaches zero.

This process shows up most clearly on the graph of distance versus time, Fig. 9-8. There, the 15 mph average speed is shown as a dotted line from 0 to the 2-hour time points. The angle this line makes with the time axis (called its *slope*) is directly related to the average speed, as we noted a few paragraphs earlier.

We can shorten the time interval from two hours to anything we like, and connect the points of the solid line which are defined by this new interval, with another angled *rate* line. Extending the rate line to intersect the time axis, then measuring the resulting slope, gives us the average speed over the new time interval (Fig. 9-9).

As we make the interval ever shorter, the *rate* line becomes ever harder to construct, but the principle remains the same. When we make the interval so short that the rate line only touches the speed curve, rather than crossing it, we have found the limit of the rate of change of distance, at that point.

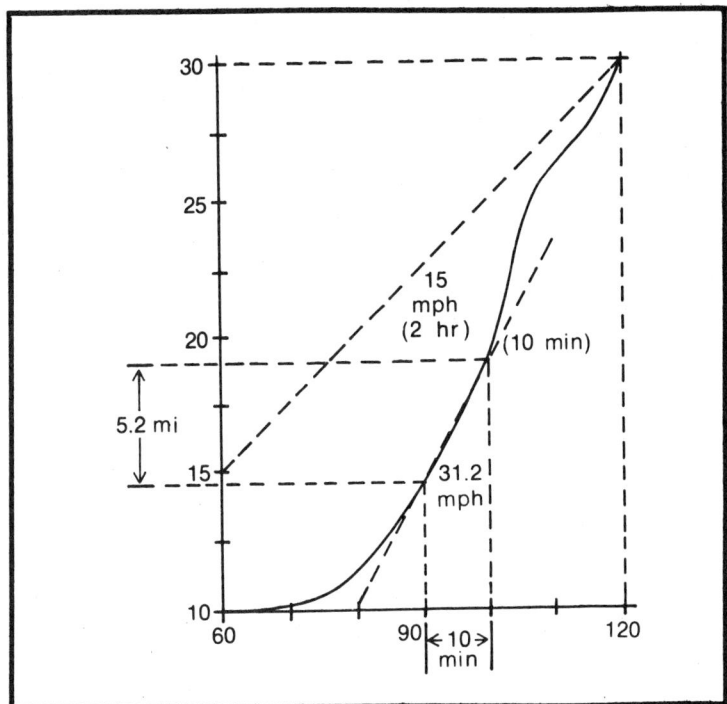

Fig. 9-9. This is an expanded view of the last hour of the trip shown in Fig. 9-8. Note that **average speed** depends upon time interval over which average is taken.

It is important to note that speed cannot change instantaneously. That is, it takes a definite length of time to change the positional rate of change of a moving object such as an automobile.

This means that the limit of the rate of change of distance at one point is not necessarily the same as that at another. If we plotted these limits on another graph (Fig. 9-10), we would find that the limits themselves have limits.

The process of determining the limit of a rate of change is called obtaining the derivative or *differentiating* the expression. The whole study of differential calculus involves rules for doing so, without having to draw a graph as we have done for illustration.

The Derivative

In earlier chapters we traced in detail the manner in which counting led to addition tables, and addition led to mul-

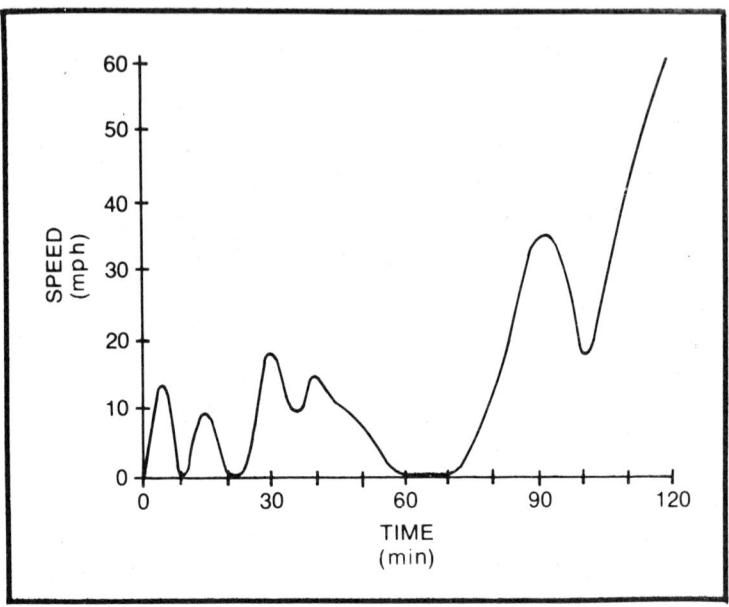

Fig. 9-10. If speedometer readings, rather than distance covered, had been plotted, the chart might have looked like this.

tiplication tables. We then followed this process through to higher orders of arithmetic operations.

Similarly, differentiation follows a progression from its basic root, up to a set of rules which can be memorized. First, let's look at that basic root, which is called by mathematicians the *delta* process of differentiation.

To apply the delta process, we must first state the relationship in the form of an equation of dependent and independent variables. That is, we must express y as a function of x, where x represents the independent variable and y represents the dependent one, and calculate the value of y for the desired value of x.

Next, we must increment x by some small quantity, and determine what effect this has upon the value of y. The new value of x is called x *plus delta* x or $x + \Delta x$, and the new value of y is called y *plus delta* y, or $y + \Delta y$, to distinguish them from the original values.

The third step of the delta process is to subtract the new values obtained in step two, from the originals obtained in step one. This produces an expression giving us a value for Δy as

224

the difference between the value of the function of $x + \Delta x$ and the value of the function x.

When we have this value for Δy, we divide the entire equation by Δx, doing the division on both sides of the equals sign to maintain equality. This gives us a ratio $\Delta y : \Delta x$ on the left side of the equation, and a rather complicated mathematical expression on the right.

The increment which we applied to x back in step two, to obtain the value of $x + \Delta x$, can be of any convenient size. It is useful, in many cases, to try various values of increments, and repeat this process many times, to get an idea of how the function behaves.

However, to obtain the derivative of the function we must determine the limit of the ratio of Δy to Δx as the value of Δx approaches zero, just as we found the limiting slope on our graph.

This means that we must use ever smaller values for Δx, until we reach a value at which the value of the ratio does not change appreciably. When we have reached this point, the value of the ratio Δy to Δx is the derivative of the function.

It may not be obvious, but the process must be capable of being carried through all five steps in order to work. This means that the limit must be reachable in the fifth step. This requirement is not met by all functions in which we might be interested. If you find yourself unable to complete step five, don't be discouraged—you may be working with one of the functions which simply cannot be differentiated.

As one differential calculus textbook puts it, "The student should be wary of functions that behave in an unorthodox fashion." The author goes on to observe that functions that jump about a lot and whose actions are quite disconnected may be, like people with similar symptoms, pathological.

Now that we have gone through the delta process of differentiation, we are in a position to appreciate some of the shortcut rules which achieves the same result without all the computation, just as multiplication achieves the same result as a much more tedious exercise in counting.

The first rule of differentiation is this: *The derivative of a constant is zero.* That is, if your function does not involve an independent variable, but consists entirely of constants, no change is possible; therefore, the rate of change must be zero.

The second rule of differentiation is equally simple: *The derivative of the independent variable, with respect to itself, is*

1. That is, when you change the independent variable, you change it by whatever amount you change it. No modification is introduced.

The third rule of differentiation begins to move away from the more obvious points: *In a function of the form $y = ax$, the derivative of y with respect to x is equal to the coefficient a.* That is, if the relationship between the dependent variable and the independent variable is a constant ratio, the dependent variable will change by that ratio, for any change in the independent variable.

Thus, if $y = 3x$, every unit change in x will cause three units change in the value of y. Similarly, if $y = x/10$, every unit change in x will produce 1/10th unit change in the value of y.

The fourth rule of differentiation begins to get more involved than any of the preceding ones. This rule involves functions in which the dependent variable is some power of the independent variable, such as the equation $y = x^n$.

For such functions, the rule, simplified, states that the derivative of y with respect to x (written dy/dx) is equal to $nx^{(n-1)}$. That is, the exponent becomes a coefficient, and the new exponent is the previous one *reduced by one*. If the equation already included a coefficient, the new coefficient is the old one multiplied by the old exponent.

This fourth rule can stand illustration. First, we have to keep firmly in mind that any value for a derivative is valid only for the immediate region around the point at which the derivative was obtained. That is, in our imaginary automobile trip, the fact that we were doing 60 mph two hours after the start says nothing about our speed at the one-hour point (which was zero), but is a close approximation to our speed at the 119-minute and 121-minute points.

With this restraint in mind, let's see how the area of a square varies when we change the length of its sides. The formula (or equation) for the area of a square is $A = L^2$. The rule tells us that the derivative of A with respect to L is $2L$.

Let's check this. If the square is four feet on a side, it will have an area of 4^2, or 16, square feet. If the derivative of the area is $2L$, its value will be 2×4, or 8, for the 4-foot square. This means that (for small changes) each fractional change in the side length will change the area by 8 times that fraction.

If we increase the side length of our square from 4 to 4.1 feet, the fractional increase is 0.1 foot. The derivative then predicts an increase in the area of the square of 8×0.1, or 0.8 square foot.

The actual area of the 4.1-foot square is 16.81 square feet, which is a 0.81 square foot increase. The prediction was off by only 1/100 square foot, an error of considerably less than 1%. For changes which are less than 1% (our example was a change of 1 part in 40, or 2.5%), the error will be much smaller.

Our fourth rule is actually a general form of which rule three is a special case. If we have a function $Y = AX^n$, rule four tells us the derivative dy/dx will be $nAX^{(n-1)}$. If n is equal to 1, the original function is $Y = AX^1$, which is the same as $Y = AX$, and the derivative is given as $1 \times A \times X$, the same as the $A \times X$ given by rule three.

As a matter of fact, rule one also is only another special case of rule four. The constant, in rule one, is the same as AX^0, for which rule four gives a derivative of $0 \times A$ or 0.

Thus our four rules are really only two, and one of these (rule two) does not involve a dependent variable. Therefore all of these rules boil down to a single elementary function involved with differentiation.

Amazingly enough, this single function will serve many of the purposes for which differentiation is used. Problems of motion, velocity, acceleration, area, and volume can all be treated with no more information than you already have at this point.

Differential calculus includes 21 more fundamental elementary functions, for which similar special rules exist. This is not the place to go into them in detail; if you are interested, a number of college textbooks discuss the rules in detail. You have enough background knowledge on the subject by now to be able to use these texts.

So What Good Is It?

Now that we have discovered what a derivative is and how to calculate it by one of the most general rules for differentiation, what can we use if for? This question did not arise with arithmetic or algebra, because their uses fall into the everyday world of almost anyone. The differential calculus, however, deals with slightly more exotic subject matter, and thus merits an examination of its applications.

One wide application of differentiation is to locate the points at which any mathematical function capable of being differentiated reaches either a maximum or a minimum.

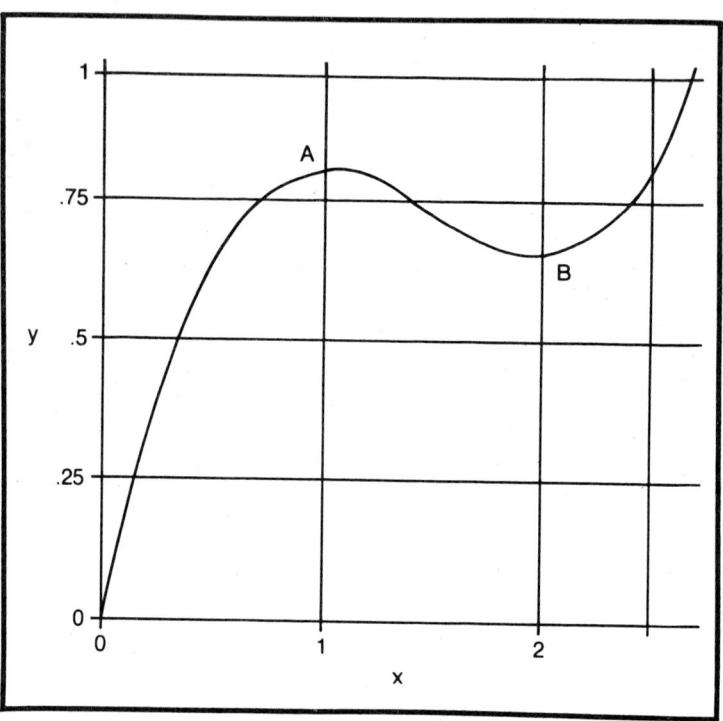

Fig. 9-11. Differential calculus can be used to determine exact values for X at both the maximum **A** and minimum **B** shown on this partial graph of function **Y = X³3 − 3X²2 + 2X**.

Figure 9-11 shows what we mean. As the value of X increases, starting from 0, the value of Y rises to a peak (labeled A), then falls until it bottoms out at the point labeled B, somewhere near the value $X = 2$. What we want to know are the exact values of X at the highest point A and the lowest point B shown on the graph.

If we know the equation which corresponds to the curve, we can use the rules for differentiation to determine the points at which the derivative reaches zero. These will include every point at which the direction of the change in Y reverses, as it must do at either a maximum or a minimum value.

For the direction of change to reverse, the rate of change must first decrease to zero, then pass through zero; for this reason, the points at which the rate of change equals zero indicate the points at which the change reverses.

As it happens, the specific curve shown is the graph of the equation $Y = X^3/3 - 3X^2/2 + 2X$. Applying the rule for differentiation, we determine that dy/dx is equal to $X^2 - 3X + 2$.

Now all we need do is solve this algebraic equation for the points at which its value is equal to zero. We can begin by factoring the equation into $(X - 2)(X - 1) = 0$, then setting each part in turn equal to zero, and we find that the *critical values* at which Y equals zero are $X = 1$ and $X = 2$.

We can tell whether the critical points found represent maxima or minima by varying the values of X slightly in both directions from the critical values and determining the direction in which the derivative moves away from zero.

For instance, a value for X of 0.99 gives dy/dx equal to 0.0101, and a value of 1.01 for X makes dy/dx equal to -0.0099. This shows us that the point $X = 1$ is a maximum for the function; the value of Y at this point is 5/6.

At the point $X = 2$, we find that reducing the value of X to 1.99 makes dy/dx negative, at -0.0099. Increasing X to 2.01 drives dy/dx positive, to 0.0100999, so this critical point is a minimum.

That is, if dy/dx becomes less than zero when X is reduced, and greater than zero when X is increased, we are at a minimum. Vice versa indicates that we are at a maximum. If the sign of dy/dx does not change when we pass through the critical point, we are at neither a maximum nor a minimum, but rather at a plateau of the function, like the coffee stop on our auto trip.

While every maximum or minimum is at a critical point, you see, not all critical points represent maxima or minima. This means that every one must be tested to determine whether it is a maximum, a minimum, or a plateau.

Let's try another example of this. Scientists found, years ago, that the force of gravity is equivalent to an acceleration of 32 feet per second per second (which they frequently write as $32 \, ft/sec^2$). That is, if you drop something from a high tower (and ignore air resistance), its velocity one second later is 32 ft/sec. After two seconds, velocity is 64 ft/sec, after three seconds it is 96 ft/sec, and so on. The longer it falls, the faster it goes.

The distance the object covers during its fall is related to both velocity and time. The average speed over any interval is half of the speed at the end of the interval, so for the first second the average would be one half of 32 ft/sec, or 16 ft/sec, making the distance covered in the first second 16 ft. Similarly, the average speed over the first two seconds is one half of 64, or 32 ft/sec, and at this speed the object moves 64 ft in the two-second period.

Over the first three seconds, the average speed is 48 ft/sec (by a similar calculation), and an object going this fast for three seconds would cover 144 feet.

We can tabulate these figures for as long a time as we like. Figure 9-12 shows them for the first 40 seconds of an object's fall.

If we study the tabulation in Fig. 9-12, we will find that the distance covered in every case is equal to $16t^2$, where the variable t indicates the time in seconds. Thus, at two seconds the distance is $16 \times 2 \times 2$, or 64 feet. At 10 seconds, it is $16 \times 10 \times 10$, or 1600. At 25 seconds, the value is $16 \times 25 \times 25$, or 10,000, and so forth.

We can also discover a pattern for velocity. The velocity is always $32t$, where t, as before, indicates time in seconds. Ten seconds after the start velocity is 320 ft/sec, at 40 seconds it is 1180 ft/sec.

You may remember that distance and time are basic physical units, while velocity (speed) is the derivative of distance with respect to time (dd/dt), and acceleration is dv/dt (dd^2/dt). If we hold acceleration constant at 32 ft/sec^2, we would expect the velocity to increase every second by 32 ft/sec—or $32t$—and distance to change as indicated by the change in velocity—$(32t/2)t$, which boils down to $16t^2$.

Looking at it from a different point of view, the derivative of the function $d = 16t^2$ is $32t$, which checks with the change of velocity. The derivative of the function $v = 32t$ is a constant value 32, which checks with the force of gravity.

Now that we see how distance and velocity change with the passage of time, so long as an object is in free fall and unaffected by anything other than the force of gravity, let's use this data to determine how high a .45-caliber bullet from a Colt automatic pistol would climb, how long it would take to get there, when it would get back down, and how fast it would be moving when it hit the ground.

The actual muzzle velocity of .45 ammunition ranges from 900 to 1200 ft/sec. We'll use a value of 1056 ft/sec because it makes the arithmetic a little easier.

When we fire the gun straight up toward the sky, the cartridge will shove the bullet out the barrel at 1056 ft/sec. If the force of gravity were not affecting it, the bullet might go on forever at this same speed. During the first second, it would climb 1056 ft. During the next, another 1056 would be covered, for a total of 2212 ft, and so forth. The slug would keep rising forever, and never return to the ground.

However, from the instant the bullet leaves the barrel, gravity is pulling against it just exactly as it would accelerate

TIME	DISTANCE	VELOCITY
0	0	0
1	16	32
2	64	64
3	144	96
4	256	128
5	400	160
6	576	192
7	784	224
8	1024	256
9	1296	288
10	1600	320
11	1936	352
12	2304	384
13	2704	416
14	3136	448
15	3600	480
16	4096	512
17	4624	544
18	5184	576
19	5776	608
20	6400	640
21	7056	672
22	7744	704
23	8464	736
24	9216	768
25	10000	800
26	10816	832
27	11664	864
28	12544	896
29	13456	920
30	14400	960
31	15376	992
32	16384	1024
33	17424	1056
34	18496	1088
35	19600	1120
36	20736	1152
37	21904	1184
38	23104	1216
39	24336	1248
40	25600	1280

Fig. 9-12. **Distance** in feet, and **velocity** in feet per second, after times from 0 to 40 seconds in free fall. Distances and velocities listed here ignore effect of air resistance.

the bullet in free fall from a height. Thus, after one second, the bullet's velocity would be reduced to 1056−32, or 1024 ft/sec. After 10 seconds, it would be down to 1056−320, or 736 ft/sec. And after 33 seconds the velocity would be 1056−1056, or 0.

Since velocity is the derivative of distance with respect to time, the fact that the velocity falls to 0 tells us that the bullet is at its maximum distance from the ground. At this point, we know that it takes 33 seconds for the force of gravity to neutralize the muzzle velocity of the cartridge, but we do not yet know how far the bullet climbed in this time.

If gravity had no effect on it, in 33 seconds the bullet would have climbed 33 times 1056, or 34,848 feet. The distance that the bullet would have fallen during that same 33 seconds is given by our function $d = 16t^2$, as $16 \times 33 \times 33$, or 17,424 ft. Since, in actual fact, the force of gravity was dragging the bullet back down at the same time that the muzzle velocity was driving it up, the net distance traveled would be the difference between these two figures—34,848−17,424, or 17,424 ft. That's about three and one half miles into the air.

To state the same thing differently, we know the initial velocity was 1056 ft/sec, and 33 seconds later the velocity was 0 ft/sec. Since the loss of velocity occurred at a uniform rate of 32 ft/sec, the average velocity over the 33 seconds is 1056/2, or 528 ft/sec. In 33 seconds the bullet climbs 17,424 ft.

Let's sum up what we have calculated so far in this problem. We fired a bullet into the air at a muzzle velocity of 1056 ft/sec, and we know that 33 seconds later it will be at its maximum height, 17,424 ft. At this instant, the velocity of the bullet is 0.

If you were in the right place at the right time, you could reach out and pick it up without injury, since it would be, quite literally, standing still in the air.

We still need to find out how long it will take the bullet to get back to earth from its 17,424 ft altitude, and how fast it will be moving when it hits.

Actually, we already have that information in our tabulation of velocity and distance for the free-fall situation, since the bullet is falling with no force other than the force of gravity applied, from a height of 17,424 feet.

One second after it begins to fall, its velocity will be 32 ft/sec. Two seconds after it reaches maximum altitude, the velocity will be 64 ft/sec, and so on, just as we discussed earlier.

The formulas tell us that another 33 seconds must pass before the 17,424 ft altitude has been traversed by the falling bullet, and at the end of that time, it will be traveling exactly as fast as it was when it left the barrel of the pistol: 1056 feet per second.

Unless you happen to be a gun nut, these velocities in feet per second may not have much meaning to you. Almost everyone is familiar with speed in miles per hour, however, so let's translate some of the figures.

A speed of 60 miles per hour is the same as 88 ft/sec, so we can convert feet per second into miles per hour by multiplying the "ft/sec" value by the ratio 60/88. When we do so, we find that the muzzle velocity of the .45-caliber bullet is 720 mph—just a little slower than the speed of sound.

Five seconds after the bullet leaves the gun, its velocity has been reduced by gravity to 896 ft/sec, which converts to 610.909 mph—still fast enough to do considerable damage to anything it might hit. At the 29-second point, when the bullet has climbed to an altitude greater than 17,150 ft, its speed is still appreciable—87.3 mph.

As we just saw, the falling bullet regains all the velocity it lost to gravity on its climb, and is traveling faster than 700 mph when it slams into the ground 66 seconds after being fired from the gun.

These facts may explain why gun enthusiasts take a dim view of anyone firing warning shots into the air under any circumstances. It is much safer to everyone in the area for the warning shot to be fired into the ground: then it cannot fall back to earth.

Since the equations we developed along the way describe the bullet's position at every instant during its flight, it is not difficult to calculate the velocity and distance of travel for every second during the 1.1 minutes of the bullets's flight period. Figure 9-13 shows such a listing.

The manner in which the velocity changes at a constant rate as time passes is shown by the graph in Fig. 9-14. This graph shows the increase in velocity during free-fall conditions, together with the decrease in velocity from the 1056 ft/sec figure. You can see that at any specific time, the two velocity values total up to the original 1056 ft/sec value.

The vertical travel of the bullet, in feet, is shown by Fig. 9-15. In this graph, the x axis represents time, with the $t = 0$

TIME (sec)	GRAVITATIONAL DISTANCE (feet)	GRAVITATIONAL VELOCITY (ft/sec)	BULLET DISTANCE	BULLET VELOCITY
0	0	0	0	1056
1	16	32	1040	1056
2	64	64	2048	992
3	144	96	3024	960
4	256	128	3968	928
5	400	160	4880	896
6	576	192	5760	864
7	784	224	6608	832
8	1024	256	7424	800
9	1296	288	8208	768
10	1600	320	8960	736
11	1936	352		
12	2304	384		
13	2704	416		
14	3136	448		
15	3600	480	8960	576
20	6400	640	12240	416
25	10000	800	14720	256
30	14400	960	17280	96
31	15376	992	17360	64
32	16384	1024	17408	32
33	17424	1056	17424	0

Fig. 9-13. Effects of gravity on climbing bullet for muzzle velocity of 1056 ft/sec, ignoring air resistance.

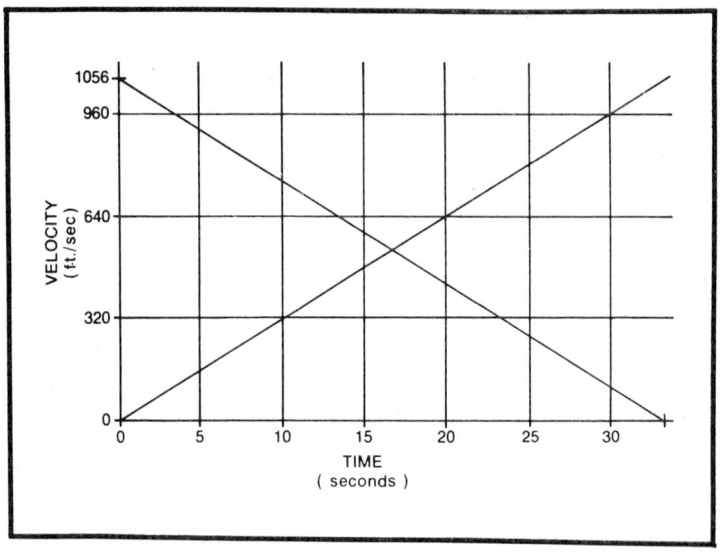

Fig. 9-14. How rise of velocity due to gravity cancels initial velocity of bullet.

corresponding to the instant at which the bullet is fired. The y axis of the graph represents vertical distance from the ground. You can see that the bullet rises in accordance with exactly the same laws it follows during its fall. The only effect of the initial velocity is to change the height achieved, and the time required to get there.

For instance, if we cut the charge of powder in the cartridge so that the muzzle velocity is just half what we have been using, or 528 ft/sec, the slug would travel for only 33 seconds total before hitting the ground, and would be moving at 528 ft/sec when it struck. At the halfway point of its flight, 16½ seconds after being fired from the pistol, it would reach a maximum altitude of 4356 ft. That is, the flight time would be half as great, and the altitude one quarter that reached by the faster bullet.

If instead of changing the muzzle velocity of the cartridge, we took the whole thing to the moon, where the force of gravity is approximately $5\frac{1}{3}$ ft/sec rather than 32, we would find a much greater altitude and longer flight time resulting.

Instead of 33 seconds to reach the peak altitude, the bullet would take more than 198 seconds to get there, and instead of a

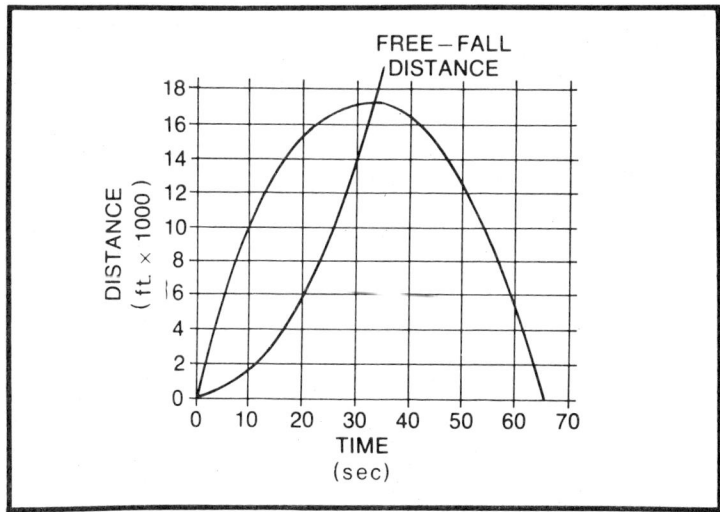

Fig. 9-15. Parabolic curve in this graph shows altitude of bullet at various times after gun is fired. Rising curve which climbs off upper side of graph represents distance object would fall, due to gravity alone, in same periods of time. Notice that gravitational curve crosses altitude curve at 33 sec mark, which indicates that gravity's force cancels bullet's upward velocity at this point.

235

measly 17,424 ft, the altitude would be 104,609. Obviously, anyone who wants to achieve maximum range from a pistol should take it to a low-gravity environment.

Of course, in such an environment the kick from the gun would have a much greater effect upon the shoulder, but that is a real-world problem, not a mathematical one.

Another application of differentials or derivatives is to determine approximate values for increments of the dependent variable in a complicated function. It may be rather tedious to compute the exact value of the increment Δy of some complicated function, for a given increment Δx of the independent variable. At the same time, it may be relatively easy to compute the differential dy/dx.

What is more, if you are dealing with actual real-world measurements, it is usually a waste of time to compute Δy exactly because the measurements themselves are only approximations and are not accurate. Thus, by computing the differential dy/dx, then multiplying this value times the increment Δx of the independent variable, we can determine a perfectly practical approximation of the value of y for the incremented function.

A similar application of the derivative is determining the error, relative error, and percentage error of measured values. Usable approximations of these three quantities are, respectively, dy, dy/y, and $100\, dy/y$.

Putting Things Back Together

So far in this chapter, we have learned how to take a function apart and determine its derivative. Now, let's see how we can determine the original function, given its derivative as a starting point.

For instance, if we are given the derivative $dy/dx = X^3$, our problem will be to find the function Y, which is some function of X whose derivative is X^3.

From what we know about how derivatives are obtained, we can see that Y must be something like X^4. However, if we assume that Y is equal to X^4, and differentiate this function, we obtain for dy/dx the value $4X^3$, which is not the derivative we were given.

The value we just computed is four times too large. This shows us what to do next—divide by four. If we state the function as $Y = X^4/4$, and differentiate this function, we get $dy/dx = X^3$, which is the derivative with which we began.

It might seem from this example that the process of finding the function, given the derivative, is at least as simple as that of finding the derivative if we are given the function. However, you must remember that the derivative of a constant is zero by our first rule of differentiation. If the original function had been, instead of $Y = X^4/4$ (as we computed it), $Y = X^4/4 + 1000$, its derivative would still have been X^3.

No matter what the constant added to the function happens to be, the derivative would not be changed. This means that the number of functions which could have the same derivative is infinite or unlimited, and therefore we have no way of putting the constant back if our only knowledge of the function is the derivative itself.

The missing constant is usually represented in textbooks by C, and is called *constant of integration*. The process of reversing differentiation (the inverse of differentiation) is known as integration. The entire technique for performing it is known as *integral calculus*.

While the process of differentiation is unique, consisting of one-to-one mapping from one set of values to another, that of integration is not. Since the constant of integration vanishes during differentiation and cannot be restored from information contained only in the derivative, all functions which differ one from another only in their constant terms are indistinguishable, once differentiated and then integrated back to original form. That is, the integration process is a one-to-many mapping.

Like the process of differentiation, integration depends upon application of a number of shortcut rules. For integration, 27 *elementary integrals* are known. The study of integral calculus is, largely, a matter of memorizing these fundamental integrals and learning how to combine them in practical applications. We won't go into that kind of detail here; we will, instead, examine the aims and purposes of the process.

For our purposes, we will use only the first rule for integration, which is the one we met at the beginning of this section—by reversing the differentiation process, we find the function which has the appropriate derivative.

The value this process gives us is called the *indefinite integral*, since the constant of integration C is not defined.

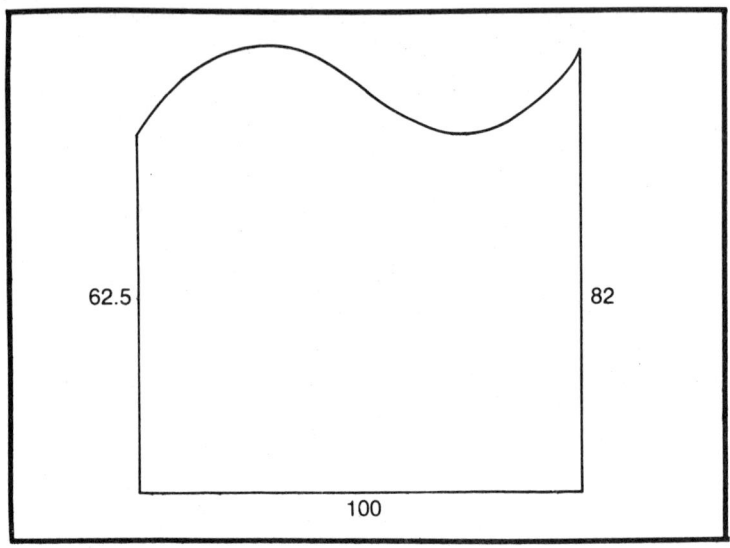

Fig. 9-16. Principles of integration can be used to determine area of this plot, wich has three straight sides and one curved one. Technique used is the basis of the integral calculus.

If, however, we know the value of the function for at least one point in its range, we can use this information in addition to the information given by the derivative, to change the integral from the *indefinite* integral into a *definite* integral, by adding or subtracting the constant as required to make the calculated value correspond with the real world at that point in the range of the function.

Another way of stating this is sometimes given as a definition for the definite integral: *The definite integral of any function of X is the difference between two values of the integral of the function, for two distinct values of the variable X.* These two values are called the *limits of integration*, and the function is said to be integrated between these limits.

One application of integration is to determine the area beneath the curve of a graph. That is, if we have a plot (Fig. 9-16) bounded on three sides by straight lines, but on the fourth by a curve that can be represented by an equation, we can determine the area of that plot (by integrating the function) between the limits represented by the two parallel straight sides.

We first met the idea of differentiation by taking an average rate of change over some time interval, then progressively slicing the time interval ever smaller, until we reached the concept of an instantaneous rate of change, or *limit*.

Since we started with a wide base and made it ever more narrow when differentiating, it's only natural to expect that we would begin with a narrow base and make it ever wider when performing the inverse operation, integration. We can see how this works, by considering the problem of integrating the area under the curve which we mentioned in the previous paragraph. Here's how:

Any curve can be approximated by a series of short, straight lines. In the length of each straight segment is sufficiently small, the error will be negligible. Our first step, then, is to convert the curve from its original smooth form into such a series of straight-line segments (Fig. 9-17).

Having done this, our next step is to connect each point at which two line segments on the former curve meet, to the base, by a line which is parallel to the existing sides of the figure. When we have completed all this line construction, we will have converted the original figure into a series of strips.

Each strip will have three sides forming the three sides of a rectangle, and the fourth side will be at an angle across the

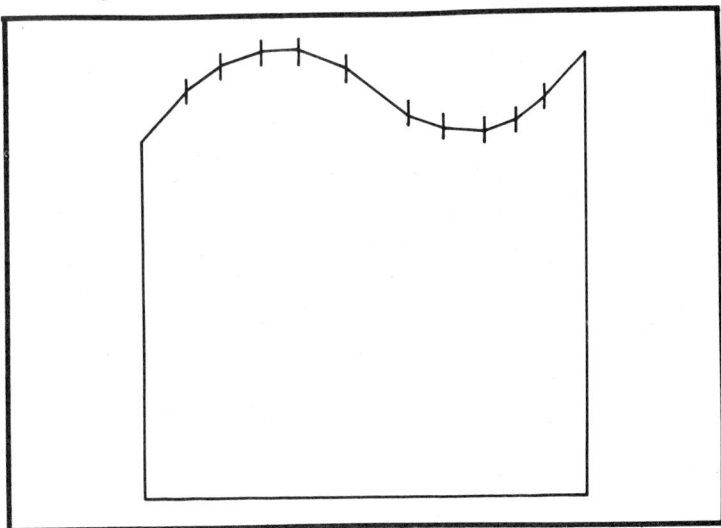

Fig. 9-17. First step in determining area is to convert smooth curve into a sequence of straight line segments. Tick marks indicate limits of each line segment. Accuracy is increased by using more segments, but we're using as few as possible in order to show the principle more clearly.

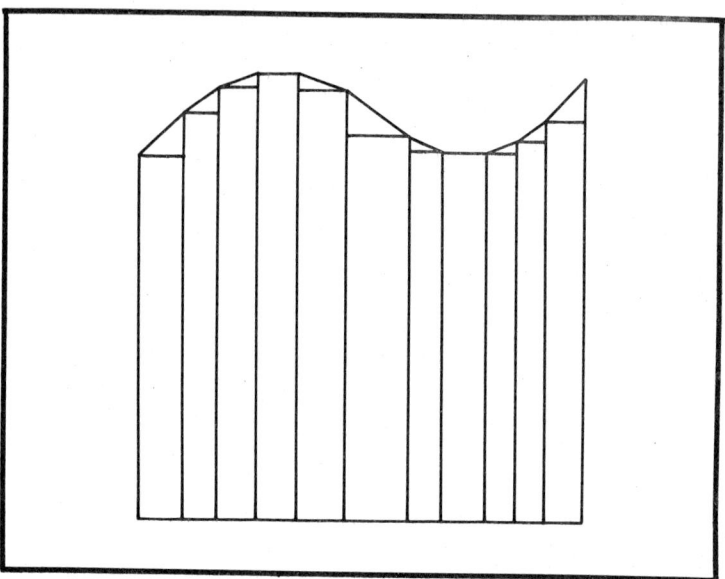

Fig. 9-18. After converting curve to line segments, vertical lines are drawn to convert figure into strips, and angled top of each strip is converted to triangle. Areas of resulting rectangles and triangles can then be found readily to provide desired figure for area of entire plot.

strip. Now, we can draw another line from the shorter side of each strip, at right angles to the side at its upper end, to cross the strip and meet the other vertical side. This converts each strip into a combination of a rectangle and a right triangle (Fig. 9-18).

The rules for calculating area tell us how to calculate the area of each rectangle (length times width), and each triangle (half of base times altitude). The total area, then, will be the sum total of all the areas we calculate; that is, the areas of all the rectangles added together, and summed with the areas of all the triangles.

The areas thus found is not exactly the same as the area under the original curve, because we introduced some error when we approximated the curve by a sequence of straight-line segments. The error can, though, be made as small as we desire, by making the line segments (and consequently the strips) as narrow as we please. This is exactly the same concept we used in differentiation, when we reduced the time interval for the averaging process to almost nothing at all.

Many other methods for integrating various expressions are known, but a study of them involves more detail than we have room for in this volume. If you are interested, you have enough background to get under way in your own independent study.

As we have presented them here, differential calculus and integral calculus have been treated as separate but related subjects of study, since that is the way U.S. educators usually present them. In Europe, they are considered to be two branches of a single subject—*infinitesimal calculus.*

Appendix A
Squares, Cubes, Square Roots, and Cube Roots

No.	Square	Cube	Square Root	Cube Root	No. = Diam.	
					Circum.	Area
1	1	1	1.0000	1.0000	3.142	0.7854
2	4	8	1.4142	1.2599	6.283	3.1416
3	9	27	1.7321	1.4423	9.425	7.0686
4	16	64	2.0000	1.5874	12.566	12.5664
5	25	125	2.2361	1.7100	15.708	19.6350
6	36	216	2.4495	1.8171	18.850	28.2743
7	49	343	2.6458	1.9129	21.991	38.4845
8	64	512	2.8284	2.0000	25.133	50.2655
9	81	729	3.0000	2.0801	28.274	63.6173
10	100	1000	3.1623	2.1544	31.416	78.5398
11	121	1331	3.3166	2.2240	34.558	95.0332
12	144	1728	3.4641	2.2894	37.699	113.097
13	169	2197	3.6056	2.3513	40.841	132.732
14	196	2744	3.7417	2.4101	43.982	153.938
15	225	3375	3.8730	2.4662	47.124	176.715
16	256	4096	4.0000	2.5198	50.265	201.062
17	289	4913	4.1231	2.5713	53.407	226.980
18	324	5832	4.2426	2.6207	56.549	254.469
19	361	6859	4.3589	2.6684	59.690	283.529
20	400	8000	4.4721	2.7144	62.832	314.159
21	441	9261	4.5826	2.7589	65.973	346.361
22	484	10648	4.6904	2.8020	69.115	380.133
23	529	12167	4.7958	2.8439	72.257	415.476
24	576	13824	4.8990	2.8845	75.398	452.389
25	625	15625	5.0000	2.9240	78.540	490.874
26	676	17576	5.0990	2.9625	81.681	530.929
27	729	19683	5.1962	3.0000	84.823	572.555
28	784	21952	5.2915	3.0366	87.965	615.752
29	841	24389	5.3852	3.0723	91.106	660.520
30	900	27000	5.4772	3.1072	94.248	706.858
31	961	29791	5.5678	3.1414	97.389	754.768
32	1024	32768	5.6569	3.1748	100.531	804.248
33	1089	35937	5.7446	3.2075	103.673	855.299
34	1156	39304	5.8310	3.2396	106.814	907.920
35	1225	42875	5.9161	3.2711	109.956	962.113
36	1296	46656	6.0000	3.3019	113.097	1017.88
37	1369	50653	6.0828	3.3322	116.239	1075.21
38	1444	54872	6.1644	3.3620	119.381	1134.11
39	1521	59319	6.2450	3.3912	122.522	1194.59

No.	Square	Cube	Square Root	Cube Root	No. = Diam. Circum.	Area
40	1600	64000	6.3246	3.4200	125.66	1256.64
41	1681	68921	6.4031	3.4482	128.81	1320.25
42	1764	74088	6.4807	3.4760	131.95	1385.44
43	1849	79507	6.5574	3.5034	135.09	1452.20
44	1936	85184	6.6332	3.5303	138.23	1520.53
45	2025	91125	6.7082	3.5569	141.37	1590.43
46	2116	97336	6.7823	3.5830	144.51	1661.90
47	2209	103823	6.8557	3.6088	147.65	1734.94
48	2304	110592	6.9282	3.6342	150.80	1809.56
49	2401	117649	7.0000	3.6593	153.94	1885.74
50	2500	125000	7.0711	3.6840	157.08	1963.50
51	2601	132651	7.1414	3.7084	160.22	2042.82
52	2704	140608	7.2111	3.8325	163.36	2123.72
53	2809	148877	7.2801	3.7563	166.50	2206.18
54	2916	157464	7.3485	3.7798	169.65	2290.22
55	3025	166375	7.4162	3.8030	172.79	2375.83
56	3136	175616	7.4833	3.8259	175.93	2463.01
57	3249	185193	7.5498	3.8485	179.07	2551.76
58	3364	195112	7.6158	3.8709	182.21	2642.08
59	3481	205379	7.6811	3.8930	185.35	2733.97
60	3600	216000	7.7460	3.9149	188.50	2827.43
61	3721	226981	7.8102	3.9365	191.64	2922.47
62	3844	238328	7.8740	3.9579	194.78	3019.07
63	3969	250047	7.9373	3.9791	197.92	3117.25
64	4096	262114	8.0000	4.0000	201.06	3216.99
65	4225	274625	8.0623	4.0207	204.20	3318.31
66	4356	287496	8.1240	4.0412	207.35	3421.19
67	4489	300763	8.1854	4.0615	210.49	3525.65
68	4624	314432	8.2462	4.0817	213.63	3631.68
69	4761	328509	8.3066	4.1016	216.77	3739.28
70	4900	343000	8.3666	4.1213	219.91	3848.45
71	5041	357911	8.4261	4.1408	223.05	3959.19
72	5184	373248	8.4853	4.1602	226.19	4071.50
73	5329	389017	8.5440	4.1793	229.34	4185.39
74	5476	405224	8.6023	4.1983	232.48	4300.84
75	5625	421875	8.6603	4.2172	235.62	4417.86
76	5776	438976	8.7178	4.2358	238.76	4536.46
77	5929	456533	8.7750	4.2543	241.90	4656.63
78	6084	474552	8.8318	4.2727	245.04	4778.36
79	6241	493039	8.8882	4.2908	248.19	4901.67

No.	Square	Cube	Square Root	Cube Root	No. = Diam.	
					Circum.	Area
80	6400	512000	8.9443	4.3089	251.33	5026.55
81	6561	531441	9.0000	4.3267	254.47	5153.00
82	6724	551368	9.0554	4.3445	257.61	5281.02
83	6889	571787	9.1104	4.3621	260.75	5410.61
84	7056	592704	9.1652	4.3795	263.89	5541.77
85	7225	614125	9.2195	4.3968	267.04	5674.50
86	7396	636056	9.2736	4.4140	270.18	5808.80
87	7569	658503	9.3274	4.4310	273.32	5944.68
88	7744	681472	9.3808	4.4480	276.46	6082.12
89	7921	704969	9.4340	4.4647	279.60	6221.14
90	8100	729000	9.4868	4.4814	282.74	6361.73
91	8281	753571	9.5394	4.4979	285.88	6503.88
92	8464	778688	9.5917	4.5144	289.03	6647.61
93	8649	804357	9.6437	4.5307	292.17	6792.91
94	8836	830584	9.6954	4.5468	295.31	6939.78
95	9025	857375	9.7468	4.5629	298.45	7088.22
96	9216	884736	9.7980	4.5789	301.59	7238.23
97	9409	912673	9.4889	4.5947	304.73	7389.81
98	9604	941192	9.8995	4.6104	307.88	7542.96
99	9801	970299	9.9499	4.6261	311.02	7697.69
100	10000	1000000	10.0000	4.6416	314.16	7853.98
101	10201	1030301	10.4099	4.6570	317.30	8011.85
102	10404	1061208	10.0995	4.6723	320.44	8171.28
103	10609	1092727	10.1489	4.6875	323.58	8332.29
104	10816	1124864	10.1980	4.7027	326.73	8494.87
105	11025	1157625	10.2470	4.7177	329.87	8659.01
106	11236	1191016	10.2956	4.7326	333.01	8824.73
107	11449	1225043	10.3441	4.7475	336.15	8992.02
108	11664	1259712	10.3923	4.7622	339.29	9160.88
109	11881	1295029	10.4403	4.7769	342.43	9331.32
110	12100	1331000	10.4881	4.7914	345.58	9503.32
111	12321	1367631	10.5357	4.8059	348.72	9676.89
112	12544	1404928	10.5830	4.8203	351.86	9852.03
113	12769	1442897	10.6301	4.8346	355.00	10028.7
114	12996	1481544	10.6771	4.8488	358.14	10207.0
115	13225	1520875	10.7238	4.8629	361.28	10386.9
116	13456	1560896	10.7703	4.8770	364.42	10568.3
117	13689	1601613	10.8167	4.8910	367.57	10751.3
118	13924	1643032	10.8628	4.9049	370.71	10935.9
119	14161	1685159	10.9087	4.9187	373.85	11122.0

No.	Square	Cube	Square Root	Cube Root	No. = Diam.	
					Circum.	Area
120	14400	1728000	10.9545	4.9324	376.99	11309.7
121	14641	1771561	11.0000	4.9461	380.13	11499.0
122	14884	1815848	11.0454	4.9597	383.27	11689.9
123	15129	1860867	11.0905	4.9732	386.42	11882.3
124	15376	1906624	11.1355	4.9866	389.56	12076.3
125	15625	1953125	11.1803	5.0000	392.70	12271.8
126	15876	2000376	11.2250	5.0133	395.84	12469.0
127	16129	2048383	11.2694	5.0265	398.98	12667.7
128	16384	2097152	11.3137	5.0397	402.12	12868.0
129	16641	2146689	11.3578	5.0528	405.27	13069.8
130	16900	2197000	11.4018	5.0658	408.41	13273.2
131	17161	2248091	11.4455	5.0788	411.55	13478.2
132	17424	2299968	11.4891	5.0916	414.69	13684.8
133	17689	2352637	11.5326	5.1045	417.83	13892.9
134	17956	2406104	11.5758	5.1172	420.97	14102.6
135	18225	2460375	11.6190	5.1299	424.12	14313.9
136	18496	2515456	11.6619	5.1426	427.26	14526.7
137	18769	2571353	11.7047	5.1551	430.40	14741.1
138	19044	2628072	11.7473	5.1676	433.54	14957.1
139	19321	2685619	11.7898	5.1801	436.68	15174.7
140	19600	2744000	11.8322	5.1925	439.82	15393.8
141	19881	2803221	11.8743	5.2048	442.96	15614.5
142	20164	2863288	11.9164	5.2171	446.11	15836.8
143	20449	2924207	11.9583	5.2293	449.25	16060.6
144	20736	2985984	12.0000	5.2415	452.39	16286.0
145	21025	3048625	12.0416	5.2536	455.53	16513.0
146	21316	3112136	12.0830	5.2656	458.67	16741.5
147	21609	3176523	12.1244	5.2776	461.81	16971.7
148	21904	3241792	12.1655	5.2896	464.96	17203.4
149	22201	3307949	12.2066	5.3015	468.10	17436.6
150	22500	3375000	12.2474	5.3133	471.24	17671.5
151	22801	3442951	12.2882	5.3251	474.38	17907.9
152	23104	3511808	12.3288	5.3368	477.52	18145.8
153	23409	3581577	12.3693	5.3485	480.66	18385.4
154	23716	3652264	12.4097	5.3601	483.81	18626.5
155	24025	3723875	12.4499	5.3717	486.95	18869.2
156	24336	3796416	12.4900	5.3832	490.09	19113.4
157	24649	3869893	12.5300	5.3947	493.23	19359.3
158	24964	3944312	12.5698	5.4061	496.37	19606.7
159	25281	4019679	12.6095	5.4175	499.51	19855.7

No.	Square	Cube	Square Root	Cube Root	No. = Diam.	
					Circum.	Area
160	25600	4096000	12.6491	5.4288	502.65	20106.2
161	25921	4173281	12.6886	5.4401	505.80	20358.3
162	26244	4251528	12.7279	5.4514	508.94	20612.0
163	26569	4330747	12.7671	5.4626	512.08	20867.2
164	26896	4410944	12.8062	5.4737	515.22	21124.1
165	27225	4492125	12.8452	5.4848	518.36	21382.5
166	27556	4574296	12.8841	5.4959	521.50	21642.4
167	27889	4657463	12.9228	5.5069	524.65	21904.0
168	28224	4741632	12.9615	5.5178	527.79	22167.1
169	28561	4826809	13.0000	5.5288	530.93	22431.8
170	28900	4913000	13.0384	5.5397	534.07	22698.0
171	29241	5000211	13.0767	5.5505	537.21	22965.8
172	29584	5088448	13.1149	5.5613	540.35	23235.2
173	29929	5177717	13.1529	5.5721	543.50	23506.2
174	30276	5268024	13.1909	5.5828	546.64	23778.7
175	30625	5359375	13.2288	5.5934	549.78	24052.8
176	30976	5451776	13.2665	5.6041	552.92	24328.5
177	31329	5545233	13.3041	5.6147	556.06	24605.7
178	31684	5639752	13.3417	5.6252	559.20	24884.6
179	32041	5735339	13.3791	5.6357	562.35	25164.9
180	32400	5832000	13.4164	5.6462	565.49	25446.9
181	32761	5929741	13.4536	5.6567	568.63	25730.4
182	33124	6028568	13.4907	5.6671	571.77	26015.5
183	33489	6128487	13.5277	5.6774	574.91	26302.2
184	33856	6229504	13.5647	5.6877	578.05	26590.4
185	34225	6331625	13.6015	5.6980	581.19	26880.3
186	34596	6434856	13.6382	5.7083	584.34	27171.6
187	34969	6539203	13.6748	5.7185	587.48	27464.6
188	35344	6644672	13.7113	5.7287	590.62	27759.1
189	35721	6751269	13.7477	5.7388	593.76	28055.2
190	36100	6859000	13.7840	5.7489	596.90	28352.9
191	36481	6967871	13.8203	5.7590	600.04	28652.1
192	36864	7077888	13.8564	5.7690	603.19	28952.9
193	37249	7189057	13.8924	5.7790	606.33	29255.3
194	37636	7301384	13.9284	5.7890	609.47	29559.2
195	38025	7414875	13.9642	5.7989	612.61	29864.8
196	38416	7529536	14.0000	5.8088	615.75	30171.9
197	38809	7645373	14.0357	5.8186	618.89	30480.5
198	39204	7762392	14.0712	5.8285	622.04	30790.7
199	39601	7880599	14.1067	5.8383	625.18	31102.6

No.	Square	Cube	Square Root	Cube Root	No. = Diam.	
					Circum.	Area
200	40000	8000000	14.1421	5.8480	628.32	31415.9
201	40401	8120601	14.1774	5.8578	631.46	31730.9
202	40804	8242408	14.2127	5.8675	634.60	32047.4
203	41209	8365427	14.2478	5.8771	637.74	32365.5
204	41616	8489664	14.2829	5.8868	640.89	32685.1
205	42025	8615125	14.3178	5.8964	644.03	33006.4
206	42436	8741816	14.3527	5.9059	647.17	33329.2
207	42849	8869743	14.3875	5.9155	650.31	33653.5
208	43264	8998912	14.4222	5.9250	653.45	33979.5
209	43681	9129329	14.4568	5.9345	656.59	34307.0
210	44100	9261000	14.4914	5.9439	659.73	34636.1
211	44521	9393931	14.5258	5.9533	662.88	34966.7
212	44944	9528128	14.5602	5.9627	666.02	35298.9
213	45369	9663597	14.5945	5.9721	669.16	35632.7
214	45796	9800344	14.6287	5.9814	672.30	35968.1
215	46225	9938375	14.6629	5.9907	675.44	36305.0
216	46656	10077696	14.6969	6.0000	678.58	36643.5
217	47089	10218313	14.7309	6.0092	681.73	36983.6
218	47524	10360232	14.7648	6.0185	684.87	37325.3
219	47961	10503459	14.7986	6.0277	688.01	37668.5
220	48400	10648000	14.8324	6.0368	691.15	38013.3
221	48841	10793861	14.8661	6.0459	694.29	38359.6
222	49284	10941048	14.8997	6.0550	697.43	38707.6
223	49729	11089567	14.9332	6.0641	700.58	39057.1
224	50176	11239424	14.9666	6.0732	703.72	39408.1
225	50625	11390625	15.0000	6.0822	706.86	39760.8
226	51076	11543176	15.0333	6.0912	710.00	40115.0
227	51529	11697083	15.0665	6.1002	713.14	40470.8
228	51984	11852352	15.0997	6.1091	716.28	40828.1
229	52441	12008989	15.1327	6.1180	719.42	41187.1
230	52900	12167000	15.1658	6.1269	722.57	41547.6
231	53361	12326391	15.1987	6.1358	725.71	41909.6
232	53824	12487168	15.2315	6.1446	728.85	42273.3
233	54289	12649337	15.2643	6.1534	731.99	42638.5
234	54756	12812904	15.2971	6.1622	735.13	43005.3
235	55225	12977875	15.3297	6.1710	738.27	43373.6
236	55696	13144256	15.3623	6.1797	741.42	43743.5
237	56169	13312053	15.3948	6.1885	744.56	44115.0
238	56644	13481272	15.4272	6.1972	747.70	44488.1
239	57121	13651919	15.4596	6.2058	750.84	44862.7

No.	Square	Cube	Square Root	Cube Root	No. = Diam.	
					Circum.	Area
240	57600	13824000	15.4919	6.2145	753.98	45238.9
241	58081	13997521	15.5242	6.2231	757.12	45616.7
242	58564	14172488	15.5563	6.2317	760.27	45996.1
243	59049	14348907	15.5885	6.2403	763.41	46377.0
244	59536	14526784	15.6205	6.2488	766.55	46759.5
245	60025	14706125	15.6525	6.2573	769.69	47143.5
246	60516	14886936	15.6844	6.2658	772.83	47529.2
247	61009	15069223	15.7162	6.2743	775.97	47916.4
248	61504	15252992	15.7480	6.2828	779.12	48305.1
249	62001	15438249	15.7797	6.2912	782.26	48695.5
250	62500	15625000	15.8114	6.2996	785.40	49087.4
251	63001	15813251	15.8430	6.3080	788.54	49480.9
252	63504	16003008	15.8745	6.3164	791.68	49875.9
253	64009	16194277	15.9060	6.3247	794.82	50272.6
254	64516	16387064	15.9374	6.3330	797.96	50670.7
255	65025	16581375	15.9687	6.3413	801.11	51070.5
256	65536	16777216	16.0000	6.3496	804.25	51471.9
257	66049	16974593	16.0312	6.3579	807.39	51874.8
258	66564	17173512	16.0624	6.3661	810.53	52279.2
259	67081	17373979	16.0935	6.3743	813.67	52685.3
260	67600	17576000	16.1245	6.3825	816.81	53092.9
261	68121	17779581	16.1555	6.3907	819.96	53502.1
262	68644	17984728	16.1864	6.3988	823.10	53912.9
263	69169	18191447	16.2173	6.4070	826.24	54325.2
264	69696	18399744	16.2481	6.4151	829.38	54739.1
265	70225	18609625	16.2788	6.4232	832.52	55154.6
266	70756	18821096	16.3095	6.4312	835.66	55571.6
267	71289	19034163	16.3401	6.4393	838.81	55990.3
268	71824	19248832	16.3707	6.4473	841.95	56410.4
269	72361	19465109	16.4012	6.4553	845.09	56832.2
270	72900	19683000	16.4317	6.4633	848.23	57255.5
271	73441	19902511	16.4621	6.4713	851.37	57680.4
272	73984	20123648	16.4924	6.4792	854.51	58106.9
273	74529	20346417	16.5227	6.4872	857.66	58534.9
274	75076	20570824	16.5529	6.4951	860.80	58964.6
275	75625	20796875	16.5831	6.5030	863.94	59395.7
276	76176	21024576	16.6132	6.5108	867.08	59828.5
277	76729	21253933	16.6433	6.5187	870.22	60262.8
278	77284	21484952	16.6733	6.5265	873.36	60698.7
279	77841	21717639	16.7033	6.5343	876.50	61136.2

No.	Square	Cube	Square Root	Cube Root	No. = Diam.	
					Circum.	Area
280	78400	21952000	16.7332	6.5421	879.65	61575.2
281	78961	22188041	16.7631	6.5499	882.79	62015.8
282	79524	22425768	16.7929	6.5577	885.93	62458.0
283	80089	22665187	16.8226	6.5654	889.07	62901.8
284	80656	22906304	16.8523	6.5731	892.21	63347.1
285	81225	23149125	16.8819	6.5808	895.35	63794.0
286	81796	23393656	16.9115	6.5885	898.50	64242.4
287	82369	23639903	16.9411	6.5962	901.64	64692.5
288	82944	23887872	16.9706	6.6039	904.78	65144.1
289	83521	24137569	17.0000	6.6115	907.92	65597.2
290	84100	24389000	17.0294	6.6191	911.06	66052.0
291	84681	24642171	17.0587	6.6267	914.20	66508.3
292	85264	24897088	17.0880	6.6343	917.35	66966.2
293	85849	25153757	17.1172	6.6419	920.49	67425.6
294	86436	25412184	17.1464	6.6494	923.63	67886.7
295	87025	25672375	17.1756	6.6569	926.77	68349.3
296	87616	25934336	17.2047	6.6644	929.91	68813.5
297	88209	26198073	17.2337	6.6719	933.05	69279.2
298	88804	26463592	17.2627	6.6794	936.19	69746.5
299	89401	26730899	17.2916	6.6869	939.34	70215.4
300	90000	27000000	17.3205	6.6943	942.48	70685.8
301	90601	27270901	17.3494	6.7018	945.62	71157.9
302	91204	27543608	17.3781	6.7092	948.76	71631.5
303	91809	27818127	17.4069	6.7166	951.90	72106.6
304	92416	28094464	17.4356	6.7240	955.04	72583.4
305	93025	28372625	17.4642	6.7313	958.19	73061.7
306	93636	28652616	17.4929	6.7387	961.33	73541.5
307	94249	28934443	17.5214	6.7460	964.47	74023.0
308	94864	29218112	17.5499	6.7533	967.61	74506.0
309	95481	29503629	17.5784	6.7606	970.75	74990.6
310	96100	29791000	17.6068	6.7679	973.89	75476.8
311	96721	30080231	17.6352	6.7752	977.04	75964.5
312	97344	30371328	17.6635	6.7824	980.18	76453.8
313	97969	30664297	17.6918	6.7897	983.32	76944.7
314	98596	30959144	17.7200	6.7969	986.46	77437.1
315	99225	31255875	17.7482	6.8041	989.60	77931.1
316	99856	31554496	17.7764	6.8113	992.74	78426.7
317	100489	31855013	17.8045	6.8185	995.88	78923.9
318	101124	32157432	17.8326	6.8256	999.03	79422.6
319	101761	32461759	17.8606	6.8328	1002.2	79922.9

No.	Square	Cube	Square Root	Cube Root	No. = Diam.	
					Circum.	Area
320	102400	32768000	17.8885	6.8399	1005.3	80424.8
321	103041	33076161	17.9165	6.8470	1008.5	80928.2
322	103684	33386248	17.9444	6.8541	1011.6	81433.2
323	104329	33698267	17.9722	6.8612	1014.7	81939.8
324	104976	34012224	18.0000	6.8683	1017.9	82448.0
325	105625	34328125	18.0278	6.8753	1021.0	82957.7
326	106276	34645976	18.0555	6.8824	1024.2	83469.0
327	106929	34965783	18.0831	6.8894	1027.3	83981.8
328	107584	35287552	18.1108	6.8964	1030.4	84496.3
329	108241	35611289	18.1384	6.9034	1033.6	85012.3
330	108900	35937000	18.1659	6.9104	1036.7	85529.9
331	109561	36264691	18.1934	6.9174	1039.9	86049.0
332	110224	36594368	18.2209	6.9244	1043.0	86569.7
333	110889	36926037	18.2483	6.9313	1046.2	87092.0
334	111556	37259704	18.2757	6.9382	1049.3	87615.9
335	112225	37595375	18.3030	6.9451	1052.4	88141.3
336	112896	37933056	18.3303	6.9521	1055.6	88668.3
337	113569	38272753	18.3576	6.9589	1058.7	89196.9
338	114244	38614472	18.3848	6.9658	1061.9	89727.0
339	114921	38958219	18.4120	6.9727	1065.0	90258.7
340	115600	39304000	18.4391	6.9795	1068.1	90792.0
341	116281	39651821	18.4662	6.9864	1071.3	91326.9
342	116964	40001688	18.4932	6.9932	1074.4	91863.3
343	117649	40353607	18.5203	7.0000	1077.6	92401.3
344	118336	40707584	18.5472	7.0068	1080.7	92940.9
345	119025	41063625	18.5742	7.0136	1083.8	93482.0
346	119716	41421736	18.6011	7.0203	1087.0	94024.7
347	120409	41781923	18.6279	7.0271	1090.1	94569.0
348	121104	42144192	18.6548	7.0338	1093.3	95114.9
349	121801	42508549	18.6815	7.0406	1096.4	95662.3
350	122500	42875000	18.7083	7.0473	1099.6	96211.3
351	123201	43243551	18.7350	7.0540	1102.7	96761.8
352	123904	43614208	18.7617	7.0607	1105.8	97314.0
353	124609	43986977	18.7883	7.0674	1109.0	97867.7
354	125316	44361864	18.8149	7.0740	1112.1	98423.0
355	126025	44738875	18.8414	7.0807	1115.3	98979.8
356	126736	45118016	18.8680	7.0873	1118.4	99538.2
357	127449	45499293	18.8944	7.0940	1121.5	100098
358	128164	45882712	18.9209	7.1006	1124.7	100660
359	128881	46268279	18.9473	7.1072	1127.8	101223

No.	Square	Cube	Square Root	Cube Root	No. Circum.	Diam. Area
360	129600	46656000	18.9737	7.1138	1131.0	101788
361	130321	47045881	19.0000	7.1204	1134.1	102354
362	131044	47437928	19.0263	7.1269	1137.3	102922
363	131769	47832147	19.0526	7.1335	1140.4	103491
364	132496	48228544	19.0788	7.1400	1143.5	104062
365	133225	48627125	19.1050	7.1466	1146.7	104635
366	133956	49027896	19.1311	7.1531	1149.8	105209
367	134689	49430863	19.1572	7.1596	1153.0	105785
368	135424	49836032	19.1833	7.1661	1156.1	106362
369	136161	50243409	19.2094	7.1726	1159.2	106941
370	136900	50653000	19.2354	7.1791	1162.4	107521
371	137641	51064811	19.2614	7.1855	1165.5	108103
372	138384	51478848	19.2873	7.1920	1168.7	108687
373	139129	51895117	19.3132	7.1984	1171.8	109272
374	139876	52313624	19.3391	7.2048	1175.0	109858
375	140625	52734375	19.3649	7.2112	1178.1	110447
376	141376	53157376	19.3907	7.2177	1181.2	111036
377	142129	53582633	19.4165	7.2240	1184.4	111628
378	142884	54010152	19.4422	7.2304	1187.5	112221
379	143641	54439939	19.4679	7.2368	1190.7	112815
380	144400	54872000	19.4936	7.2432	1193.8	113411
381	145161	55306341	19.5192	7.2495	1196.9	114009
382	145924	55742968	19.5448	7.2558	1200.1	114608
383	146689	56181887	19.5704	7.2622	1203.2	115209
384	147456	56623104	19.5959	7.2685	1206.4	115812
385	148225	57066625	19.6214	7.2748	1209.5	116416
386	148996	57512456	19.6469	7.2811	1212.7	117021
387	149769	57960603	19.6723	7.2874	1215.8	117628
388	150544	58411072	19.6977	7.2936	1218.9	118237
389	151321	58863869	19.7231	7.2999	1222.1	118847
390	152100	59319000	19.7484	7.3061	1225.2	119459
391	152881	59776471	19.7737	7.3124	1228.4	120072
392	153664	60236288	19.7990	7.3186	1231.5	120687
393	154449	60698457	19.8242	7.3248	1234.6	121304
394	155236	61162984	19.8494	7.3310	1237.8	121922
395	156025	61629875	19.8746	7.3372	1240.9	122542
396	156816	62099136	19.8997	7.3434	1244.1	123163
397	157609	62570773	19.9249	7.3496	1247.2	123786
398	158404	63044792	19.9499	7.3558	1250.4	124410
399	159201	63521199	19.9750	7.3619	1253.5	125036

No.	Square	Cube	Square Root	Cube Root	No. = Diam.	
					Circum.	Area
400	160000	64000000	20.0000	7.3681	1256.6	125664
401	160801	64481201	20.0250	7.3742	1259.8	126293
402	161604	64964808	20.0499	7.3803	1262.9	126923
403	162409	65450827	20.0749	7.3864	1266.1	127556
404	163216	65939264	20.0998	7.3925	1269.2	128190
405	164025	66430125	20.1246	7.3986	1272.3	128825
406	164836	66923416	20.1494	7.4047	1275.5	129462
407	165649	67419143	20.1742	7.4108	1278.6	130100
408	166464	67917312	20.1990	7.4169	1281.8	130741
409	167281	68417929	20.2237	7.4229	1284.9	131382
410	168100	68921000	20.2485	7.4290	1288.1	132025
411	168921	69426531	20.2731	7.4350	1291.2	132670
412	169744	69934528	20.2978	7.4410	1294.3	133317
413	170569	70444997	20.3224	7.4470	1297.5	133965
414	171396	70957944	20.3470	7.4530	1300.6	134614
415	172225	71473375	20.3715	7.4590	1303.8	135265
416	173056	71991296	20.3961	7.4650	1306.9	135918
417	173889	72511713	20.4206	7.4710	1310.0	136572
418	174724	73034632	20.4450	7.4770	1313.2	137228
419	175561	73560059	20.4695	7.4829	1316.3	137885
420	176400	74088000	20.4939	7.4889	1319.5	138544
421	177241	74618461	20.5183	7.4948	1322.6	139205
422	178084	75151448	20.5426	7.5007	1325.8	139867
423	178929	75686967	20.5670	7.5067	1328.9	140531
424	179776	76225024	20.5913	7.5126	1332.0	141196
425	180625	76765625	20.6155	7.5185	1335.2	141863
426	181476	77308776	20.6398	7.5244	1338.3	142531
427	182329	77854483	20.6640	7.5302	1341.5	143201
428	183184	8402752	20.6882	7.5361	1344.6	143872
429	184041	78953589	20.7123	7.5420	1347.7	144545
430	184900	79507000	20.7364	7.5478	1350.9	145220
431	185761	80062991	20.7605	7.5537	1354.0	145896
432	186624	80621568	20.7846	7.5595	1357.2	146574
433	187489	81182737	20.8087	7.5654	1360.3	147254
434	188356	81746504	20.8327	7.5712	1363.5	147934
435	189225	82312875	20.8567	7.5770	1366.6	148617
436	190096	82881856	20.8806	7.5828	1369.7	149301
437	190969	83453453	20.9045	7.5886	1372.9	149987
438	191844	84027672	20.9284	7.5944	1376.0	150674
439	192721	84604519	20.9523	7.6001	1379.2	151363

No.	Square	Cube	Square Root	Cube Root	No. = Diam.	
					Circum.	Area
440	193600	85184000	20.9762	7.6059	1382.3	152053
441	194481	85766121	21.0000	7.6117	1385.4	152745
442	195364	86350888	21.0238	7.6174	1388.6	153439
443	196249	86938307	21.0476	7.6232	1391.7	154134
444	197136	87528384	21.0713	7.6289	1394.9	154830
445	198025	88121125	21.0950	7.6346	1398.0	155528
446	198916	88716536	21.1187	7.6403	1401.2	156228
447	199809	89314623	21.1424	7.6460	1404.3	156930
448	200704	89915392	21.1660	7.6517	1407.4	157633
449	201601	90518849	21.1896	7.6574	1410.6	158337
450	202500	91125000	21.2132	7.6631	1413.7	159043
451	203401	91733851	21.2368	7.6688	1416.9	159751
452	204304	92345408	21.2603	7.6744	1420.0	160460
453	205209	92959677	21.2838	7.6801	1423.1	161171
454	206116	93576664	21.3073	7.6857	1426.3	161883
455	207025	94196375	21.3307	7.6914	1429.4	162597
456	207936	94818816	21.3542	7.6970	1432.6	163313
457	208849	95443993	21.3776	7.7026	1435.7	164030
458	209764	96071912	21.4009	7.7082	1438.9	164748
459	210681	96702579	21.4243	7.7138	1442.0	165468
460	211600	97336000	21.4476	7.7194	1445.1	166190
461	212521	97972181	21.4709	7.7250	1448.3	166914
462	213444	98611128	21.4942	7.7306	1451.4	167639
463	214369	99252847	21.5174	7.7362	1454.6	168365
464	215296	99897344	21.5407	7.7418	1457.7	169093
465	216225	100544625	21.5639	7.7473	1460.8	169823
466	217156	101194696	21.5870	7.7529	1464.0	170554
467	218089	101847563	21.6102	7.7584	1467.1	171287
468	219024	102503232	21.6333	7.7639	1470.3	172021
469	219961	103161709	21.6564	7.7695	1473.4	172757
470	220900	103823000	21.6795	7.7750	1476.5	173494
471	221841	104487111	21.7025	7.7805	1479.7	174234
472	222784	105154048	21.7256	7.7860	1482.8	174974
473	223729	105823817	21.7486	7.7915	1486.0	175716
474	224676	106496424	21.7715	7.7970	1489.1	176460
475	225625	107171875	21.7945	7.8025	1492.3	177205
476	226576	107850176	21.8174	7.8079	1495.4	177952
477	227529	108531333	21.8403	7.8134	1498.5	178701
478	228484	109215352	21.8632	7.8188	1501.7	179451
479	229441	109902239	21.8861	7.8243	1504.8	180203

No.	Square	Cube	Square Root	Cube Root	No. = Diam.	
					Circum.	Area
480	230400	110592000	21.9089	7.8297	1508.0	180956
481	231361	111284641	21.9317	7.8352	1511.1	181711
482	232324	111980168	21.9545	7.8406	1514.3	182467
483	233289	112678587	21.9773	7.8460	1517.4	183225
484	234256	113379904	22.0000	7.8514	1520.5	183984
485	235225	114084125	22.0227	7.8568	1523.7	184745
486	236196	114791256	22.0454	7.8622	1526.8	185508
487	237169	115501303	22.0681	7.8676	1530.0	186272
488	238144	116214272	22.0907	7.8730	1533.1	187038
489	239121	116930169	22.1133	7.8784	1536.2	187805
490	240100	117649000	22.1359	7.8837	1539.4	188574
491	241081	118370771	22.1585	7.8891	1542.5	189345
492	242064	119095488	22.1811	7.8944	1545.7	190117
493	243049	119823157	22.2036	7.8998	1548.8	190890
494	244036	120553784	22.2261	7.9051	1551.9	191665
495	245025	121287375	22.2486	7.9105	1555.1	192442
496	246016	122023936	22.2711	7.9158	1558.2	193221
497	247009	122763473	22.2935	7.9211	1561.4	194000
498	248004	123505992	22.3159	7.9264	1564.5	194782
499	249001	124251499	22.3383	7.9317	1567.7	195565
500	250000	125000000	22.3607	7.9370	1570.8	196350
501	251001	125751501	22.3830	7.9423	1573.9	197136
502	252004	126506008	22.4054	7.9476	1577.1	197923
503	253009	127263527	22.4277	7.9528	1580.2	198713
504	254016	128024064	22.4499	7.9581	1583.4	199504
505	255025	128787625	22.4722	7.9634	1586.5	200296
506	256036	129554216	22.4944	7.9686	1589.7	201090
507	257049	130323843	22.5167	7.9739	1592.8	201886
508	258064	131096512	22.5389	7.9791	1595.9	202683
509	259081	131872229	22.5610	7.9843	1599.1	203482
510	260100	132651000	22.5832	7.9896	1602.2	204282
511	261121	133432831	22.6053	7.9948	1605.4	205084
512	262144	134217728	22.6274	8.0000	1608.5	205887
513	263169	135005697	22.6495	8.0052	1611.6	206692
514	264196	135796744	22.6716	8.0104	1614.8	207499
515	265225	136590875	22.6936	8.0156	1617.9	208307
516	266256	137388096	22.7156	8.0208	1621.1	209117
517	267289	138188413	22.7376	8.0260	1624.2	209928
518	268324	138991832	22.7596	8.0311	1627.3	210741
519	269361	139798359	22.7816	8.0363	1630.5	211556

No.	Square	Cube	Square Root	Cube Root	No. = Diam.	
					Circum.	Area
520	270400	140608000	22.8035	8.0415	1633.6	212372
521	271441	141420761	22.8254	8.0466	1636.8	213189
522	272484	142236648	22.8473	8.0517	1639.9	214008
523	273529	143055667	22.8692	8.0569	1643.1	214829
524	274576	143877824	22.8910	8.0620	1646.2	215651
525	275625	144703125	22.9129	8.0671	1649.3	216475
526	276676	145531576	22.9347	8.0723	1652.5	217301
527	277729	146363183	22.9565	8.0774	1655.6	218128
528	278784	147197952	22.9783	8.0825	1658.8	218956
529	279841	148035889	23.0000	8.0876	1661.9	219787
530	280900	148877000	23.0217	8.0927	1665.0	220618
531	281961	149721291	23.0434	8.0978	1668.2	221452
532	283024	150568768	23.0651	8.1028	1671.3	222287
533	284089	151419437	23.0868	8.1079	1674.5	223123
534	285156	152273304	23.1084	8.1130	1677.6	223961
535	286225	153130375	23.1301	8.1180	1680.8	224801
536	287296	153990656	23.1517	8.1231	1683.9	225642
537	288369	154854153	23.1733	8.1281	1687.0	226484
538	289444	155720872	23.1948	8.1332	1690.2	227329
539	290521	156590819	23.2164	8.1382	1693.3	228175
540	291600	157464000	23.2379	8.1433	1696.5	229022
541	292681	158340421	23.2594	8.1483	1699.6	229871
542	293764	159220088	23.2809	8.1533	1702.7	230722
543	294849	160103007	23.3024	8.1583	1705.9	231574
544	295936	160989184	23.3238	8.1633	1709.0	232428
545	297025	161878625	23.3452	8.1683	1712.2	233283
546	298116	162771336	23.3666	8.1733	1715.3	234140
547	299209	163667323	23.3880	8.1783	1718.5	234998
548	300304	164566592	23.4094	8.1833	1721.6	235858
549	301401	165469149	23.4307	8.1882	1724.7	236720
550	302500	166375000	23.4521	8.1932	1727.9	237583
551	303601	167284151	23.4734	8.1982	1731.0	238448
552	304704	168196608	23.4947	8.2031	1734.2	239314
553	305809	169112377	23.5160	8.2081	1737.3	240182
554	306916	170031464	23.5372	8.2130	1740.4	241051
555	308025	170953875	23.5584	8.2180	1743.6	241922
556	309136	171879616	23.5797	8.2229	1746.7	242795
557	310249	172808693	23.6008	8.2278	1749.9	243669
558	311364	173741112	23.6220	8.2327	1753.0	244545
559	312481	174676879	23.6432	8.2377	1756.2	245422

No.	Square	Cube	Square Root	Cube Root	No. = Diam.	
					Circum.	Area
560	313600	175616000	23.6643	8.2426	1759.3	246301
561	314721	176558481	23.6854	8.2475	1762.4	247181
562	315844	177504328	23.7065	8.2524	1765.6	248063
563	316969	178453547	23.7276	8.2573	1768.7	248947
564	318096	179406144	23.7487	8.2621	1771.9	249832
565	319225	180362125	23.7697	8.2670	1775.0	250719
566	320356	181321496	23.7908	8.2719	1778.1	251607
567	321489	182284263	23.8118	8.2768	1781.3	252497
568	322624	183250432	23.8328	8.2816	1784.4	253388
569	323761	184220009	23.8537	8.2865	1787.6	254281
570	324900	185193000	23.8747	8.2913	1790.7	255176
571	326041	186169411	23.8956	8.2962	1793.9	256072
572	327184	187149248	23.9165	8.3010	1797.0	256970
573	328329	188132517	23.9374	8.3059	1800.1	257869
574	329476	189119224	23.9583	8.3107	1803.3	258770
575	330625	190109375	23.9792	8.3155	1806.4	259672
576	331776	191102976	24.0000	8.3203	1809.6	260576
577	332929	192100033	24.0208	8.3251	1812.7	261482
578	334084	193100552	24.0416	8.3300	1815.8	262389
579	335241	194104539	24.0624	8.3348	1819.0	263298
580	336400	195112000	24.0832	8.3396	1822.1	264208
581	337561	196122941	24.1039	8.3443	1825.3	265120
582	338724	197137368	24.1247	8.3491	1828.4	266033
583	339889	198155287	24.1454	8.3539	1831.6	266948
584	341056	199176704	24.1661	8.3587	1834.7	267865
585	342225	200201625	24.1868	8.3634	1837.8	268783
586	343396	201230056	24.2074	8.3682	1841.0	269701
587	344569	202262003	24.2281	8.3730	1844.1	270624
588	345744	203297472	24.2487	8.3777	1847.3	271547
589	346921	204336469	24.2693	8.3825	1850.4	272471
590	348100	205379000	24.2899	8.3872	1853.5	273397
591	349281	206425071	24.3105	8.3919	1856.7	274325
592	350464	207474688	24.3311	8.3967	1859.8	275254
593	351649	208527857	24.3516	8.4014	1863.0	276184
594	352836	209584584	24.3721	8.4061	1866.1	277117
595	354025	210644875	24.3926	8.4108	1869.3	278051
596	355216	211708736	24.4131	8.4155	1872.4	278986
597	356409	212776173	24.4336	8.4202	1875.5	279923
598	357604	213847192	24.4540	8.4249	1878.7	280862
599	358801	214921799	24.4745	8.4296	1881.8	281802

No.	Square	Cube	Square Root	Cube Root	No. = Diam.	
					Circum.	Area
600	360000	216000000	24.4949	8.4343	1885.0	282743
601	361201	217081801	24.5153	8.4390	1888.1	283687
602	362404	218167208	24.5357	8.4437	1891.2	284631
603	363609	219256227	24.5561	8.4484	1894.4	285578
604	364816	220348864	24.5764	8.4530	1897.5	286526
605	366025	221445125	24.5967	8.4577	1900.7	287475
606	367236	222545016	24.6171	8.4623	1903.8	288426
607	368449	223648543	24.6374	8.4670	1907.0	289379
608	369664	224755712	24.6577	8.4716	1910.1	290333
609	370881	225866529	24.6779	8.4763	1913.2	291289
610	372100	226981000	24.6982	8.4809	1916.4	292247
611	373321	228099131	24.7184	8.4856	1919.5	293206
612	374544	229220928	24.7386	8.4902	1922.7	294166
613	375769	230346397	24.7588	8.4948	1925.8	295128
614	376996	231475544	24.7790	8.4994	1928.9	296092
615	378225	232608375	24.7992	8.5040	1932.1	297057
616	379456	233744896	24.8193	8.5086	1935.2	298024
617	380689	234885113	24.8395	8.5132	1938.4	298992
618	381924	236029032	24.8596	8.5178	1941.5	299962
619	383161	237176659	24.8797	8.5224	1944.7	300934
620	384400	238328000	24.8998	8.5270	1947.8	301907
621	385641	239483061	24.9199	8.5316	1950.9	302882
622	386884	240641848	24.9399	8.5462	1954.1	303858
623	388129	241804367	24.9600	8.5408	1957.2	304836
624	389376	242970624	24.9800	8.5453	1960.4	305815
625	390625	244140625	25.0000	8.5499	1963.5	306796
626	391876	245314376	25.0200	8.5544	1966.6	307779
627	393129	246491883	25.0400	8.5590	1969.8	308763
628	394384	247673152	25.0599	8.5635	1972.9	309748
629	395641	248858189	25.0799	8.5681	1976.1	310736
630	396900	250047000	25.0998	8.5726	1979.2	311725
631	398161	251239591	25.1197	8.5772	1982.4	312715
632	399424	252435968	25.1396	8.5817	1985.5	313707
633	400689	253636137	25.1595	8.5862	1988.6	314700
634	401956	254840104	25.1794	8.5907	1991.8	315696
635	403225	256047875	25.1992	8.5952	1994.9	316692
636	404496	257259456	25.2190	8.5997	1998.1	317690
637	405769	258474853	25.2389	8.6043	2001.2	318690
638	407044	259694072	24.2587	8.6088	2004.3	319692
639	408321	260917119	25.2784	8.6132	2007.5	320695

No.	Square	Cube	Square Root	Cube Root	No. = Diam.	
					Circum.	Area
640	409600	262144000	25.2982	8.6177	2010.6	321699
641	410881	263374721	25.3180	8.6222	2013.8	322705
642	412164	264609288	25.3377	8.6267	2016.9	323713
643	413449	265847707	25.3574	8.6312	2020.0	324722
644	414736	267089984	25.3772	8.6357	2023.2	325733
645	416025	268336125	25.3969	8.6401	2026.3	326745
646	417316	269586136	25.4165	8.6446	2029.5	327759
647	418609	270840023	25.4362	8.6490	2032.6	328775
648	419904	272097792	25.4558	8.6535	2035.8	329792
649	421201	273359449	25.4755	8.6579	2038.9	330810
650	422500	274625000	25.4951	8.6624	2042.0	331831
651	423801	275894451	25.5147	8.6668	2045.2	332853
652	425104	277167808	25.5343	8.6713	2048.3	333876
653	426409	278445077	25.5539	8.6757	2051.5	334901
654	427716	279726264	25.5734	8.6801	2054.6	335927
655	429025	281011375	25.5930	8.6845	2057.7	336955
656	430336	282300416	25.6125	8.6890	2060.9	337985
657	431649	283593393	25.6320	8.6934	2064.0	339016
658	432964	284890312	25.6515	8.6978	2067.2	340049
659	434281	286191179	25.6710	8.7022	2070.3	341084
660	435600	287496000	25.6905	8.7066	2073.5	342119
661	436921	288804781	25.7099	8.7110	2076.6	343157
662	438244	290117528	25.7294	8.7154	2079.7	344196
663	439569	291434247	25.7488	8.7198	2082.9	345237
664	440896	292754944	25.7682	8.7241	2086.0	346279
665	442225	294079625	25.7876	8.7285	2089.2	347323
666	443556	295408296	25.8070	8.7329	2092.3	348368
667	444889	296740963	25.8263	8.7373	2095.4	349415
668	446224	298077632	25.8457	8.7416	2098.6	350464
669	447561	299418309	25.8650	8.7460	2101.7	351514
670	448900	300763000	25.8844	8.7503	2104.9	352565
671	450241	302111711	25.9037	8.7547	2108.0	353618
672	451584	303464448	25.9230	8.7590	2111.2	354673
673	452929	304821217	25.9422	8.7634	2114.3	355730
674	454276	306182024	25.9615	8.7677	2117.4	356788
675	455625	307546875	25.9808	8.7721	2120.6	357847
676	456976	308915776	26.0000	8.7764	2123.7	358908
677	458329	310288733	26.0192	8.7807	2126.9	359971
678	459684	311665752	26.0384	8.7850	2130.0	361035
679	461041	313046839	26.0576	8.7893	2133.1	362101

No.	Square	Cube	Square Root	Cube Root	No. = Diam.	
					Circum.	Area
680	462400	314432000	26.0768	8.7937	2136.3	363168
681	463761	315821241	26.0960	8.7980	2139.4	364237
682	465124	317214568	26.1151	8.8023	2142.6	365308
683	466489	318611987	26.1343	8.8066	2145.7	366380
684	467856	320013504	26.1534	8.8109	2148.9	367453
685	469225	321419125	26.1725	8.8152	2152.0	368528
686	470596	322828856	26.1916	8.8194	2155.1	369605
687	471969	324242703	26.2107	8.8237	2158.3	370684
688	473344	325660672	26.2298	8.8280	2161.4	371764
689	474721	327082769	26.2488	8.8323	2164.6	372845
690	476100	328509000	26.2679	8.8366	2167.7	373928
691	477481	329939371	26.2869	8.8408	2170.8	375013
692	478864	331373888	26.3059	8.8451	2174.0	376099
693	480249	332812557	26.3249	8.8493	2177.1	377187
694	481636	334255384	26.3439	8.8536	2180.3	378276
695	483025	335702375	26.3629	8.8578	2183.4	379367
696	484416	337153536	26.3818	8.8621	2186.6	380459
697	485809	338608873	26.4008	8.8663	2189.7	381554
698	487204	340068392	26.4197	8.8706	2192.8	382649
699	488601	341532099	26.4386	8.8748	2196.0	383746
700	490000	343000000	26.4575	8.8790	2199.1	384845
701	491401	344472101	26.4764	8.8833	2202.3	385945
702	492804	345948408	26.4953	8.8875	2205.4	387047
703	494209	347428927	26.5141	8.8917	2208.5	388151
704	495616	348913664	26.5330	8.8959	2211.7	389256
705	497025	350402625	26.5518	8.9001	2214.8	390363
706	498436	351895816	26.5707	8.9043	2218.0	391471
707	499849	353393243	26.5895	8.9085	2221.1	392580
708	501264	354894912	26.6083	8.9127	2224.3	393692
709	502681	356400829	26.6271	8.9169	2227.4	394805
710	504100	357911000	26.6458	8.9211	2230.5	395919
711	505521	359425431	26.6646	8.9253	2233.7	397035
712	506944	360944128	26.6833	8.9295	2236.8	398153
713	508369	362467097	26.7021	8.9337	2240.0	399272
714	509796	363994344	26.7208	8.9378	2243.1	400393
715	511225	365525875	26.7395	8.9420	2246.2	401515
716	512656	367061696	26.7582	8.9462	2249.4	402639
717	514089	368601813	26.7769	8.9503	2252.5	403765
718	515524	370146232	26.7955	8.9545	2255.7	404892
719	516961	371694959	26.8142	8.9587	2258.8	406020

No.	Square	Cube	Square Root	Cube Root	No. = Diam.	
					Circum.	Area
720	518400	373248000	26.8328	8.9628	2261.9	407150
721	519841	374805361	26.8514	8.9670	2265.1	408282
722	521284	376367048	26.8701	8.9711	2268.2	409416
723	522729	377933067	26.8887	8.9752	2271.4	410550
724	524176	379503424	26.9072	8.9794	2274.5	411687
725	525625	381078125	26.9258	8.9835	2277.7	412825
726	527076	382657176	26.9444	8.9876	2280.8	413965
727	528529	384240583	26.9629	8.9918	2283.9	415106
728	529984	385828352	26.9815	8.9959	2287.1	416248
729	531441	387420489	27.0000	9.0000	2290.2	417393
730	532900	389017000	27.0185	9.0041	2293.4	418539
731	534361	390617891	27.0370	9.0082	2296.5	419686
732	535824	392223168	27.0555	9.0123	2299.7	420835
733	537289	393832837	27.0740	9.0164	2302.8	421986
734	538756	395446904	27.0924	9.0205	2305.9	423138
735	540225	397065375	27.1109	9.0246	2309.1	424293
736	541696	398688256	27.1293	9.0287	2312.2	425448
737	543169	400315553	27.1477	9.0328	2315.4	426604
738	544644	401947272	27.1662	9.0369	2318.5	427762
739	546121	403583419	27.1846	9.0410	2321.6	428922
740	547600	405224000	27.2029	9.0450	2324.8	430084
741	549081	406869021	27.2213	9.0491	2327.9	431247
742	550564	408518488	27.2397	9.0532	2331.1	432412
743	552049	410172407	27.2580	9.0572	2334.2	433578
744	553536	411830784	27.2764	9.0613	2337.3	434746
745	555025	413493625	27.2947	9.0654	2340.5	435916
746	556516	415160936	27.3130	9.0694	2343.6	437087
747	558009	416832723	27.3313	9.0735	2346.8	438259
748	559504	418508992	27.3496	9.0775	2349.9	439433
749	561001	420189749	27.3679	9.0816	2353.1	440609
750	562500	421875000	27.3861	9.0856	2356.2	441786
751	564001	423564751	27.4044	9.0896	2359.3	442965
752	565504	425259008	27.4226	9.0937	2362.5	444146
753	567009	426957777	27.4408	9.0977	2365.6	445328
754	568516	428661064	27.4591	9.1017	2368.8	446511
755	570025	430368875	27.4773	9.1057	2371.9	447697
756	571536	432081216	27.4955	9.1098	2375.0	448883
757	573049	433798093	27.5136	9.1138	2378.2	450072
758	574564	435519512	27.5318	9.1178	2381.3	451262
759	576081	437245479	27.5500	9.1218	2384.5	452453

No.	Square	Cube	Square Root	Cube Root	No. = Diam.	
					Circum.	Area
760	577600	438976000	27.5681	9.1258	2387.6	453646
761	579121	440711081	27.5862	9.1298	2390.8	454841
762	580644	442450728	27.6043	9.1338	2393.9	456037
763	582169	444194947	27.6225	9.1378	2397.0	457234
764	583696	445943744	27.6405	9.1418	2400.2	458434
765	585225	447697125	27.6586	9.1458	2403.3	459635
766	586756	449455096	27.6767	9.1498	2406.5	460837
767	588289	451217663	27.6948	9.1537	2409.6	462042
768	589824	452984832	27.7128	9.1577	2412.7	463247
769	591361	454756609	27.7308	9.1617	2415.9	464454
770	592900	456533000	27.7489	9.1657	2419.0	465663
771	594441	458314011	27.7669	9.1696	2422.2	466873
772	595984	460099648	27.7849	9.1736	2425.3	468085
773	597529	461889917	27.8029	9.1775	2428.5	469298
774	599076	463684824	27.8209	9.1815	2431.6	470513
775	600625	465484375	27.8388	9.1855	2434.7	471730
776	602176	467288576	27.8568	9.1894	2437.9	472948
777	603729	469097433	27.8747	9.1933	2441.0	474168
778	605284	470910952	27.8927	9.1973	2444.2	475389
779	606841	472729139	27.9106	9.2012	2447.3	476612
780	608400	474552000	27.9285	9.2052	2450.4	477836
781	609961	476379541	27.9464	9.2091	2453.6	479062
782	611524	478211768	27.9643	9.2130	2456.7	480290
783	613089	480048687	27.9821	9.2170	2459.9	481519
784	614656	481890304	28.0000	9.2209	2463.0	482750
785	616225	483736625	28.0179	9.2248	2466.2	483982
786	617796	485587656	28.0357	9.2287	2469.3	485216
787	619369	487443403	28.0535	9.2326	2472.4	486451
788	620944	489303872	28.0713	9.2365	2475.6	487688
789	622521	491169069	28.0891	9.2404	2478.7	488927
790	624100	493039000	28.1069	9.2443	2481.9	490167
791	625681	494913671	28.1247	9.2482	2485.0	491409
792	627264	496793088	28.1425	9.2521	2488.1	492652
793	628849	498677257	28.1603	9.2560	2491.3	493897
794	630436	500566184	28.1780	9.2599	2494.4	495143
795	632025	502459875	28.1957	9.2638	2497.6	496391
796	633616	504358336	28.2135	9.2677	2500.7	497641
797	635209	506261573	28.2312	9.2716	2503.8	498892
798	636804	508169592	28.2489	9.2754	2507.0	500145
799	638401	510082399	28.2666	9.2793	2510.1	501399

No.	Square	Cube	Square Root	Cube Root	No. = Diam.	
					Circum.	Area
800	640000	512000000	28.2843	9.2832	2513.3	502655
801	641601	513922401	28.3019	9.2870	2516.4	503912
802	643204	515849608	28.3196	9.2909	2519.6	505171
803	644809	517781627	28.3373	9.2948	2522.7	506432
804	646416	519718464	28.3549	9.2986	2525.8	507694
805	648025	521660125	28.3725	9.3025	2529.0	508958
806	649636	523606616	28.3901	9.3063	2532.1	510223
807	651249	525557943	28.4077	9.3102	2535.3	511490
808	652864	527514112	28.4253	9.3140	2538.4	512758
809	654481	529475129	28.4429	9.3179	2541.5	514028
810	656100	531441000	28.4605	9.3217	2544.7	515300
811	657721	533411731	28.4781	9.3255	2547.8	516573
812	659344	535387328	28.4956	9.3294	2551.0	517848
813	660969	537367797	28.5132	9.3332	2554.1	519124
814	662596	539353144	28.5307	9.3370	2557.3	520402
815	664225	541343375	28.5482	9.3408	2560.4	521681
816	665856	543338496	28.5657	9.3447	2563.5	522962
817	667489	545338513	28.5832	9.3485	2566.7	524245
818	669124	547343432	28.6007	9.3523	2569.8	525529
819	670761	549353259	28.6182	9.3561	2573.0	526814
820	672400	551368000	28.6356	9.3599	2576.1	528102
821	674041	553387661	28.6531	9.3637	2579.2	529391
822	675684	555412248	28.6705	9.3675	2582.4	530681
823	677329	557441767	28.6880	9.3713	2585.5	531973
824	678976	559476224	28.7054	9.3751	2588.7	533267
825	680625	561515625	28.7228	9.3789	2591.8	534562
826	682276	563559976	28.7402	9.3827	2595.0	535858
827	683929	565609283	28.7576	9.3865	2598.1	537157
828	685584	567663552	28.7750	9.3902	2601.2	538456
829	687241	569722789	28.7924	9.3940	2604.4	539758
830	688900	571787000	28.8097	9.3978	2607.5	541061
831	690561	573856191	28.8271	9.4016	2610.7	542365
832	692224	575930368	28.8444	9.4053	2613.8	543671
833	693889	578009537	28.8617	9.4091	2616.9	544979
834	695556	580093704	28.8791	9.4129	2620.1	546288
835	697225	582182875	28.8964	9.4166	2623.2	547599
836	698896	584277056	28.9137	9.4204	2626.4	548912
837	700569	586376253	28.9310	9.4241	2629.5	550226
838	702244	588480472	28.9482	9.4279	2632.7	551541
839	703921	590589719	28.9655	9.4316	2635.8	552858

No.	Square	Cube	Square Root	Cube Root	No. = Diam.	
					Circum.	Area
840	705600	592704000	28.9828	9.4354	2638.9	554177
841	707281	594823321	29.0000	9.4391	2642.1	555497
842	708964	596947688	29.0172	9.4429	2645.2	556819
843	710649	599077107	29.0345	9.4466	2648.4	558142
844	712336	601211584	29.0517	9.4503	2651.5	559467
845	714025	603351125	29.0689	9.4541	2654.6	560794
846	715716	605495736	29.0861	9.4578	2657.8	562122
847	717409	607645423	29.1033	9.4615	2660.9	563452
848	719104	609800192	29.1204	9.4652	2664.1	564783
849	720801	611960049	29.1376	9.4690	2667.2	566116
850	722500	614125000	29.1548	9.4727	2670.4	567450
851	724201	616295051	29.1719	9.4764	2673.5	568786
852	725904	618470208	29.1890	9.4801	2676.6	570124
853	727609	620650477	29.2062	9.4838	2679.8	571463
854	729316	622835864	29.2233	9.4875	2682.9	572803
855	731025	625026375	29.2404	9.4912	2686.1	574146
856	732736	627222016	29.2575	9.4949	2689.2	575490
857	734449	629422793	29.2746	9.4986	2692.3	576835
858	736164	631628712	29.2916	9.5023	2695.5	578182
859	737881	633839779	29.3087	9.5060	2698.6	579530
860	739600	636056000	29.3258	9.5097	2701.8	580880
861	741321	638277381	29.3428	9.5134	2704.9	582232
862	743044	640503928	29.3598	9.5171	2708.1	583585
863	744769	642735647	29.3769	9.5207	2711.2	584940
864	746496	644972544	29.3939	9.5244	2714.3	586297
865	748225	647214625	29.4109	9.5281	2717.5	587655
866	749956	649461896	29.4279	9.5317	2720.6	589014
867	751689	651714363	29.4449	9.5354	2723.8	590375
868	753424	653972032	29.4618	9.5391	2726.9	591738
869	755161	656234909	29.4788	9.5427	2730.0	593102
870	756900	658503000	29.4958	9.5464	2733.2	594468
871	758641	660776311	29.5127	9.5501	2736.3	595835
872	760384	663054848	29.5296	9.5537	2739.5	597204
873	762129	665338617	29.5466	9.5574	2742.6	598575
874	763876	667627624	29.5635	9.5610	2745.8	599947
875	765625	669921875	29.5804	9.5647	2748.9	601320
876	767376	672221376	29.5973	9.5683	2752.0	602696
877	769129	674526133	29.6142	9.5719	2755.2	604073
878	770884	676836152	29.6311	9.5756	2758.3	605451
879	772641	679151439	29.6479	9.5792	2761.5	606831

No.	Square	Cube	Square Root	Cube Root	No. = Diam.	
					Circum.	Area
880	774400	681472000	29.6648	9.5828	2764.6	608212
881	776161	683797841	29.6816	9.5865	2767.7	609595
882	777924	686128968	29.6985	9.5901	2770.9	610980
883	779689	688465387	29.7153	9.5937	2774.0	612366
884	781456	690807104	29.7321	9.5973	2777.2	613754
885	783225	693154125	29.7489	9.6010	2780.3	615143
886	784996	695506456	29.7658	9.6046	2783.5	616534
887	786769	697864103	29.7825	9.6082	2786.6	617927
888	788544	700227072	29.7993	9.6118	2789.7	619321
889	790321	702595369	29.8161	9.6154	2792.9	620717
890	792100	704969000	29.8329	9.6190	2796.0	622114
891	793881	707347971	29.8496	9.6226	2799.2	623513
892	795664	709732288	29.8664	9.6262	2802.3	624913
893	797449	712121957	29.8831	9.6298	2805.4	626315
894	799236	714516984	29.8998	9.6334	2808.6	627718
895	801025	716917375	29.9166	9.6370	2811.7	629124
896	802816	719323136	29.9333	9.6406	2814.9	630530
897	804609	721734273	29.9500	9.6442	2818.0	631938
898	806404	724150792	29.9666	9.6477	2821.2	633348
899	808201	726572699	29.9833	9.6513	2824.3	634760
900	810000	729000000	30.0000	9.6549	2827.4	636173
901	811801	731432701	30.0167	9.6585	2830.6	637587
902	813604	733870808	30.0333	9.6620	2833.7	639003
903	815409	736314327	30.0500	9.6656	2836.9	640421
904	817216	738763264	30.0666	9.6692	2840.0	641840
905	819025	741217625	30.0832	9.6727	2843.1	643261
906	820836	743677416	30.0998	9.6763	2846.3	644683
907	822649	746142643	30.1164	9.6799	2849.4	646107
908	824464	748613312	30.1330	9.6834	2852.6	647533
909	826281	751089429	30.1496	9.6870	2855.7	648960
910	828100	753571000	30.1662	9.6905	2858.8	650388
911	829921	756058031	30.1828	9.6941	2862.0	651818
912	831744	758550528	30.1993	9.6976	2865.1	653250
913	833569	761048497	30.2159	9.7012	2868.3	654684
914	835396	763551944	30.2324	9.7017	2871.4	656118
915	837225	766060875	30.2490	9.7082	2874.6	657555
916	839056	768575296	30.2655	9.7118	2877.7	658993
917	840889	771095213	30.2820	9.7153	2880.8	660433
918	842724	773620632	30.2985	9.7188	2884.0	661874
919	844561	776151559	30.3150	9.7224	2887.1	663317

No.	Square	Cube	Square Root	Cube Root	No. = Diam.	
					Circum.	Area
920	846400	778688000	30.3315	9.7259	2890.3	664761
921	848241	781229961	30.3480	9.7294	2893.4	666207
922	850084	783777448	30.3645	9.7329	2896.5	667654
923	851929	786330467	30.3809	9.7364	2899.7	669103
924	853776	788889024	30.3974	9.7400	2902.8	670554
925	855625	791453125	30.4138	9.7435	2906.0	672006
926	857476	794022776	30.4302	9.7470	2909.1	673460
927	859329	796597983	30.4467	9.7505	2912.3	674915
928	861184	799178752	30.4631	9.7540	2915.4	676372
929	863041	801765089	30.4795	9.7575	2918.5	677831
930	864900	804357000	30.4959	9.7610	2921.7	679291
931	866761	806954491	30.5123	9.7645	2924.8	680752
932	868624	809557568	30.5287	9.7680	2928.0	682216
933	870489	812166237	30.5450	9.7715	2931.1	683680
934	872356	814780504	30.5614	9.7750	2934.2	685147
935	874225	817400375	30.5778	9.7785	2937.4	686615
936	876096	820025856	30.5941	9.7819	2940.5	688084
937	877969	822656953	30.6105	9.7854	2943.7	689555
938	879844	825293672	30.6268	9.7889	2946.8	691028
939	881721	827936019	30.6431	9.7924	2950.0	692502
940	883600	830584000	30.6594	9.7959	2953.1	693978
941	885481	833237621	30.6757	9.7993	2956.2	695455
942	887364	835896888	30.6920	9.8028	2959.4	696934
943	889249	838561807	30.7083	9.8063	2962.5	698415
944	891136	841232384	30.7246	9.8097	2965.7	699897
945	893025	843908625	30.7409	9.8132	2968.8	701380
946	894916	846590536	30.7571	9.8167	2971.9	702865
947	896809	849278123	30.7734	9.8201	2975.1	704352
948	898704	851971392	30.7896	9.8236	2978.2	705840
949	900601	854670349	30.8058	9.8270	2981.4	707330
950	902500	857375000	30.8221	9.8305	2984.5	708822
951	904401	860085351	30.8383	9.8339	2987.7	710315
952	906304	862801408	30.8545	9.8374	2990.8	711809
953	908209	865523177	30.8707	9.8408	2993.9	713306
954	910116	868250664	30.8869	9.8443	2997.1	714803
955	912025	870983875	30.9031	9.8477	3000.2	716303
956	913936	873722816	30.9192	9.8511	3003.4	717804
957	915849	876467493	30.9354	9.8546	3006.5	719306
958	917764	879217912	30.9516	9.8580	3009.6	720810
959	919681	881974079	30.9677	9.8614	3012.8	722316

No.	Square	Cube	Square Root	Cube Root	No. = Diam.	
					Circum.	Area
960	921600	884736000	30.9839	9.8648	3015.9	723823
961	923521	887503681	31.0000	9.8683	3019.1	725332
962	925444	890277128	31.0161	9.8717	3022.2	726842
963	927369	893056347	31.0322	9.8751	3025.4	728354
964	929296	895841344	31.0483	9.8785	3028.5	729867
965	931225	898632125	31.0644	9.8819	3031.6	731382
966	933156	901428696	31.0805	9.8854	3034.8	732899
967	935089	904231063	31.0966	9.8888	3037.9	734417
968	937024	907039232	31.1127	9.8922	3041.1	735937
969	938961	909853209	31.1288	9.8956	3044.2	737458
970	940900	912673000	31.1448	9.8990	3047.3	738981
971	942841	915498611	31.1609	9.9024	3050.5	740506
972	944784	918330048	31.1769	9.9058	3053.6	742032
973	946729	921167317	31.1929	9.9092	3056.8	743559
974	948676	924010424	31.2090	9.9126	3059.9	745088
975	950625	926859375	31.2250	9.9160	3063.1	746619
976	952576	929714176	31.2410	9.9194	3066.2	748151
977	954529	932574833	31.2570	9.9227	3069.3	749685
978	956484	935441352	31.2730	9.9261	3072.5	751221
979	958441	938313739	31.2890	9.9295	3075.6	752758
980	960400	941192000	31.3050	9.9329	3078.8	754296
981	962361	944076141	31.3209	9.9363	3081.9	755837
982	964324	946966168	31.3369	9.9396	3085.0	757378
983	966289	949862087	31.3528	9.9430	3088.2	758922
984	968256	952763904	31.3688	9.9464	3091.3	760466
985	970225	955671625	31.3847	9.9497	3094.5	762013
986	972196	958585256	31.4006	9.9531	3097.6	763561
987	974169	961504803	31.4166	9.9565	3100.8	765111
988	976144	964430272	31.4325	9.9598	3103.9	766662
989	978121	967361669	31.4484	9.9632	3107.0	768214
990	980100	970299000	31.4643	9.9666	3110.2	769769
991	982081	973242271	31.4802	9.9699	3113.3	771325
992	984064	976191488	31.4960	9.9733	3116.5	772882
993	986049	979146657	31.5119	9.9766	3119.6	774441
994	988036	982107784	31.5278	9.9800	3122.7	776002
995	990025	985074875	31.5436	9.9833	3125.9	777564
996	992016	988047936	31.5595	9.9866	3129.0	779128
997	994009	991026973	31.5753	9.9900	3132.2	780693
998	996004	994011992	31.5911	9.9933	3135.3	782260
999	998001	997002999	31.6070	9.9967	3138.5	783828

Appendix B
Common Logarithms

N	0	1	2	3	4	5	6	7	8	9
50	6990	6998	7007	7016	7024	7033	7042	7050	7059	7067
51	7076	7084	7093	7101	7110	7118	7126	7135	7143	7152
52	7160	7168	7177	7185	7193	7202	7210	7218	7226	7235
53	7243	7251	7259	7267	7275	7284	7292	7300	7308	7316
54	7324	7332	7340	7348	7356	7364	7372	7380	7388	7396
55	7404	7412	7419	7427	7435	7443	7451	7459	7466	7474
56	7482	7490	7497	7505	7513	7520	7528	7536	7543	7551
57	7559	7566	7574	7582	7589	7597	7604	7612	7619	7627
58	7634	7642	7649	7657	7664	7672	7679	7686	7694	7701
59	7709	7716	7723	7731	7738	7745	7752	7760	7767	7774
60	7782	7789	7796	7803	7810	7818	7825	7832	7839	7846
61	7853	7860	7868	7875	7882	7889	7896	7903	7910	7917
62	7924	7931	7938	7945	7952	7959	7966	7973	7980	7987
63	7993	8000	8007	8014	8021	8028	8035	8041	8048	8055
64	8062	8069	8075	8082	8089	8096	8102	8109	8116	8122
65	8129	8136	8142	8149	8156	8162	8169	8176	8182	8189
66	8195	8202	8209	8215	8222	8228	8235	8241	8248	8254
67	8261	8267	8274	8280	8287	8293	8299	8306	8312	8319
68	8325	8331	8338	8344	8351	8357	8363	8370	8376	8382
69	8388	8395	8401	8407	8414	8420	8426	8432	8439	8445
70	8451	8457	8463	8470	8476	8482	8488	8494	8500	8506
71	8513	8519	8525	8531	8537	8543	8549	8555	8561	8567
72	8573	8579	8585	8591	8597	8603	8609	8615	8621	8627
73	8633	8639	8645	8651	8657	8663	8669	8675	8681	8686
74	8692	8698	8704	8710	8716	8722	8727	8733	8739	8745
75	8751	8756	8762	8768	8774	8779	8785	8791	8797	8802
76	8808	8814	8820	8825	8831	8837	8842	8848	8854	8859
77	8865	8871	8876	8882	8887	8893	8899	8904	8910	8915
78	8921	8927	8932	8938	8943	8949	8954	8960	8965	8971
79	8976	8982	8987	8993	8998	9004	9009	9015	9020	9025
80	9031	9036	9042	9047	9053	9058	9063	9069	9074	9079
81	9085	9090	9096	9101	9106	9112	9117	9122	9128	9133
82	9138	9143	9149	9154	9159	9165	9170	9175	9180	9186
83	9191	9196	9201	9206	9212	9217	9222	9227	9232	9238
84	9243	9248	9253	9258	9263	9269	9274	9279	9284	9289
85	9294	9299	9304	9309	9315	9320	9325	9330	9335	9340
86	9345	9350	9355	9360	9365	9370	9375	9380	9385	9390
87	9395	9400	9405	9410	9415	9420	9425	9430	9435	9440
88	9445	9450	9455	9460	9465	9469	9474	9479	9484	9489
89	9494	9499	9504	9509	9513	9518	9523	9528	9533	9538
90	9542	9547	9552	9557	9562	9566	9571	9576	9581	9586
91	9590	9595	9600	9605	9609	9614	9619	9624	9628	9633
92	9638	9643	9647	9652	9657	9661	9666	9671	9675	9680
93	9685	9689	9694	9699	9703	9708	9713	9717	9722	9727
94	9731	9736	9741	9745	9750	9754	9759	9763	9768	9773
95	9777	9782	9786	9791	9795	9800	9805	9809	9814	9818
96	9823	9827	9832	9836	9841	9845	9850	9854	9859	9863
97	9868	9872	9877	9881	9886	9890	9894	9899	9903	9908
98	9912	9917	9921	9926	9930	9934	9939	9943	9948	9952
99	9956	9961	9965	9969	9974	9978	9983	9987	9991	9996
100	0000	0004	0009	0013	0017	0022	0026	0030	0035	0039
N	0	1	2	3	4	5	6	7	8	9

COMMON LOGARITHMS

N	0	1	2	3	4	5	6	7	8	9
0	....	0000	3010	4771	6021	6990	7782	8451	9031	9542
1	0000	0414	0792	1139	1461	1761	2041	2304	2553	2788
2	3010	3222	3424	3617	3802	3979	4150	4314	4472	4624
3	4771	4914	5051	5185	5315	5441	5563	5682	5798	5911
4	6021	6128	6232	6335	6435	6532	6628	6721	6812	6902
5	6990	7076	7160	7243	7324	7404	7482	7559	7634	7709
6	7782	7853	7924	7993	8062	8129	8195	8261	8325	8388
7	8451	8513	8573	8633	8692	8751	8808	8865	8921	8976
8	9031	9085	9138	9191	9243	9294	9345	9395	9445	9494
9	9542	9590	9638	9685	9731	9777	9823	9868	9912	9956
10	0000	0043	0086	0128	0170	0212	0253	0294	0334	0374
11	0414	0453	0492	0531	0569	0607	0645	0682	0719	0755
12	0792	0828	0864	0899	0934	0969	1004	1038	1072	1106
13	1139	1173	1206	1239	1271	1303	1335	1367	1399	1430
14	1461	1492	1523	1553	1584	1614	1644	1673	1703	1732
15	1761	1790	1818	1847	1875	1903	1931	1959	1987	2014
16	2041	2068	2095	2122	2148	2175	2201	2227	2253	2279
17	2304	2330	2355	2380	2405	2430	2455	2480	2504	2529
18	2553	2577	2601	2625	2648	2672	2695	2718	2742	2765
19	2788	2810	2833	2856	2878	2900	2923	2945	2967	2989
20	3010	3032	3054	3075	3096	3118	3139	3160	3181	3201
21	3222	3243	3263	3284	3304	3324	3345	3365	3385	3404
22	3424	3444	3464	3483	3502	3522	3541	3560	3579	3598
23	3617	3636	3655	3674	3692	3711	3729	3747	3766	3784
24	3802	3820	3838	3856	3874	3892	3909	3927	3945	3962
25	3979	3997	4014	4031	4048	4065	4082	4099	4115	4133
26	4150	4166	4183	4200	4216	4232	4249	4265	4281	4298
27	4314	4330	4346	4362	4378	4393	4409	4425	4440	4456
28	4472	4487	4502	4518	4533	4548	4564	4579	4594	4609
29	4624	4639	4654	4669	4683	4698	4713	4728	4742	4757
30	4771	4786	4800	4814	4829	4843	4857	4871	4886	4900
31	4914	4928	4942	4955	4969	4983	4997	5011	5024	5038
32	5051	5065	5079	5092	5105	5119	5132	5145	5159	5172
33	5185	5198	5211	5224	5237	5250	5263	5276	5289	5302
34	5315	5328	5340	5353	5366	5378	5391	5403	5416	5428
35	5441	5453	5465	5478	5490	5502	5514	5527	5539	5551
36	5563	5575	5587	5599	5611	5623	5635	5647	5658	5670
37	5682	5694	5705	5717	5729	5740	5752	5763	5775	5786
38	5798	5809	5821	5832	5843	5855	5866	5877	5888	5899
39	5911	5922	5933	5944	5955	5966	5977	5988	5999	6010
40	6021	6031	6042	6053	6064	6075	6085	6096	6107	6117
41	6128	6138	6149	6160	6170	6180	6191	6201	6212	6222
42	6232	6243	6253	6263	6274	6284	6294	6304	6314	6325
43	6335	6345	6355	6365	6375	6385	6395	6405	6415	6425
44	6435	6444	6454	6464	6474	6484	6493	6503	6513	6522
45	6532	6542	6551	6561	6571	6580	6590	6599	6609	6618
46	6628	6637	6646	6656	6665	6675	6684	6693	6702	6712
47	6721	6730	6739	6749	6758	6767	6776	6785	6794	6803
48	6812	6821	6830	6839	6848	6857	6866	6875	6884	6893
49	6902	6911	6920	6928	6937	6946	6955	6964	6972	6981
50	6990	6998	7007	7016	7024	7033	7042	7050	7059	7067
N	0	1	2	3	4	5	6	7	8	9

Appendix C
Natural Logarithms

N	0.0	1.0	2.0	3.0	4.0	5.0	6.0	7.0	8.0	9.0
	....	0.0000	0.6931	1.0986	1.3863	1.6094	1.7918	1.9459	2.0794	2.1972
10	2.3026	2.3979	2.4849	2.5649	2.6391	2.7081	2.7726	2.8332	2.8904	2.9444
20	9957	3.0445	3.0910	3.1355	3.1781	3.2189	3.2581	3.2958	3.3322	3.3673
30	3.4012	4340	4657	4965	5264	5553	5835	6109	6376	6636
40	6889	7136	7377	7612	7842	8067	8286	8501	8712	8918
50	9120	9318	9512	9703	9890	4.0073	4.0254	4.0431	4.0604	4.0775
60	4.0943	4.1109	4.1271	4.1431	4.1589	1744	1897	2047	2195	2341
70	2485	2627	2767	2905	3041	3175	3307	3408	3567	3694
80	3820	3944	4067	4188	4308	4427	4543	4659	4773	4886
90	4998	5109	5218	5326	5433	5539	5643	5747	5850	5951
100	6052	6151	6250	6347	6444	6540	6634	6728	6821	6913
110	7005	7095	7185	7274	7362	7449	7536	7622	7707	7791
120	7875	7958	8040	8122	8203	8283	8363	8442	8520	8598
130	8675	8752	8828	8903	8978	9053	9127	9200	9273	9345
140	9416	9488	9558	9628	9698	9767	9836	9904	9972	5.0039
150	5.0106	5.0173	5.0239	5.0304	5.0370	5.0434	5.0499	5.0562	5.0626	0689
160	0752	0814	0876	0938	0999	1059	1120	1180	1240	1299
170	1358	1417	1475	1533	1591	1648	1705	1761	1818	1874
180	1930	1985	2040	2095	2149	2204	2257	2311	2364	2417
190	2470	2523	2575	2627	2679	2730	2781	2832	2883	2933
200	2983	3033	3083	3132	3181	3230	3279	3327	3375	3423
210	3471	3519	3566	3613	3660	3706	3753	3799	3845	3891
220	3936	3982	4027	4072	4116	4161	4205	4250	4293	4337
230	4381	4424	4467	4510	4553	4596	4638	4681	4723	4765
240	4806	4848	4889	4931	4972	5013	5053	5094	5134	5175
250	5215	5255	5294	5334	5373	5413	5452	5491	5530	5568
260	5607	5645	5683	5722	5759	5797	5835	5872	5910	5947
270	5984	6021	6058	6095	6131	6168	6204	6240	6276	6312
280	6348	6384	6419	6454	6490	6525	6560	6595	6630	6664
290	6699	6733	6768	6802	6836	6870	6904	6937	6971	7004
300	7038	7071	7104	7137	7170	7203	7236	7268	7301	7333
310	7366	7398	7430	7462	7494	7526	7557	7589	7621	7652
320	7683	7714	7746	7777	7807	7838	7869	7900	7930	7961
330	7991	8021	8051	8081	8111	8141	8171	8201	8230	8260
340	8289	8319	8348	8377	8406	8435	8464	8493	8522	8551
350	8579	8608	8636	8665	8693	8721	8749	8777	8805	8833
360	8861	8889	8916	8944	8972	8999	9026	9054	9081	9108
370	9135	9162	9189	9216	9243	9269	9296	9322	9349	9375
380	9402	9428	9454	9480	9506	9532	9558	9584	9610	9636
390	9661	9687	9713	9738	9764	9789	9814	9839	9865	9890
400	9915	9940	9965	9989	6.0014	6.0039	6.0064	6.0088	6.0113	6.0137
410	6.0162	6.0186	6.0210	6.0234	0259	0283	0307	0331	0355	0379
420	0403	0426	0450	0474	0497	0521	0544	0568	0591	0615
430	0638	0661	0684	0707	0730	0753	0776	0799	0822	0845
440	0868	0890	0913	0936	0958	0981	1003	1026	1048	1070
450	1092	1115	1137	1159	1181	1203	1225	1247	1269	1291
460	1312	1334	1356	1377	1399	1420	1442	1463	1485	1506
470	1527	1549	1570	1591	1612	1633	1654	1675	1696	1717
480	1738	1759	1779	1800	1821	1841	1862	1883	1903	1924
490	1944	1964	1985	2005	2025	2046	2066	2086	2106	2126
500	2146	2166	2186	2206	2226	2246	2265	2285	2305	2324
N	0.0	1.0	2.0	3.0	4.0	5.0	6.0	7.0	8.0	9.0

NATURAL LOGARITHMS

N	0.0	1.0	2.0	3.0	4.0	5.0	6.0	7.0	8.0	9.0
500	6.2146	6.2166	6.2186	6.2206	6.2226	6.2246	6.2265	6.2285	6.2305	6.2324
510	2344	2364	2383	2403	2422	2442	2461	2480	2500	2519
520	2538	2558	2577	2596	2615	2634	2653	2672	2691	2710
530	2729	2748	2766	2785	2804	2823	2841	2860	2879	2897
540	2916	2934	2953	2971	2989	3008	3026	3044	3063	3081
550	3099	3117	3135	3154	3172	3190	3208	3226	3244	3261
560	3279	3297	3315	3333	3351	3368	3386	3404	3421	3439
570	3456	3474	3491	3509	3526	3544	3561	3578	3596	3613
580	3630	3648	3665	3682	3699	3716	3733	3750	3767	3784
590	3801	3818	3835	3852	3869	3886	3902	3919	3936	3953
600	3969	3986	4003	4019	4036	4052	4069	4085	4102	4118
610	4135	4151	4167	4184	4200	4216	4232	4249	4265	4281
620	4297	4313	4329	4345	4362	4378	4394	4409	4425	4441
630	4457	4473	4489	4505	4520	4536	4552	4568	4583	4599
640	4615	4630	4646	4661	4677	4693	4708	4723	4739	4754
650	4770	4785	4800	4816	4831	4846	4862	4877	4892	4907
660	4922	4938	4953	4968	4983	4998	5013	5028	5043	5058
670	5073	5088	5103	5117	5132	5147	5162	5177	5191	5206
680	5221	5236	5250	5265	5280	5294	5309	5323	5338	5352
690	5367	5381	5396	5410	5425	5439	5453	5468	5482	5497
700	5511	5525	5539	5554	5568	5582	5596	5610	5624	5639
710	5653	5667	5681	5695	5709	5723	5737	5751	5765	5779
720	5793	5806	5820	5834	5848	5862	5876	5889	5903	5917
730	5930	5944	5958	5971	5985	5999	6012	6026	6039	6053
740	6067	6080	6093	6107	6120	6134	6147	6161	6174	6187
750	6201	6214	6227	6241	6254	6267	6280	6294	6307	6320
760	6333	6346	6359	6373	6386	6399	6412	6425	6438	6451
770	6464	6477	6490	6503	6516	6529	6542	6554	6567	6580
780	6593	6606	6619	6631	6644	6657	6670	6682	6695	6708
790	6720	6733	6746	6758	6771	6783	6796	6809	6821	6834
800	6846	6859	6871	6884	6896	6908	6921	6933	6946	6958
810	6970	6983	6995	7007	7020	7032	7044	7056	7069	7081
820	7093	7105	7117	7130	7142	7154	7166	7178	7190	7202
830	7214	7226	7238	7250	7262	7274	7286	7298	7310	7322
840	7334	7346	7358	7370	7382	7393	7405	7417	7429	7441
850	7452	7464	7476	7488	7499	7511	7523	7534	7546	7558
860	7569	7581	7593	7604	7616	7627	7639	7650	7662	7673
870	7685	7696	7708	7719	7731	7742	7754	7765	7776	7788
880	7799	7811	7822	7833	7845	7856	7867	7878	7890	7901
890	7912	7923	7935	7946	7957	7968	7979	7991	8002	8013
900	8024	8035	8046	8057	8068	8079	8090	8101	8112	8123
910	8134	8145	8156	8167	8178	8189	8200	8211	8222	8233
920	8244	8255	8265	8276	8287	8298	8309	8320	8330	8341
930	8352	8363	8373	8384	8395	8405	8416	8427	8437	8448
940	8459	8469	8480	8491	8501	8512	8522	8533	8544	8554
950	8565	8575	8586	8596	8607	8617	8628	8638	8648	8659
960	8669	8680	8690	8701	8711	8721	8732	8742	8752	8763
970	8773	8783	8794	8804	8814	8824	8835	8845	8855	8865
980	8876	8886	8896	8906	8916	8926	8937	8947	8957	8967
990	8977	8987	8997	9007	9017	9027	9037	9048	9058	9068
1000	9078	9088	9098	9108	9117	9127	9137	9147	9157	9167
N	0.0	1.0	2.0	3.0	4.0	5.0	6.0	7.0	8.0	9.0

Appendix D
Values of ϵ^x and ϵ^{-x}

x	Function	0.00	0.01	0.02	0.03	0.04	0.05	0.06	0.07	0.08	0.09
0.0	ϵ^x	1.0000	1.0101	1.0202	1.0305	1.0408	1.0513	1.0618	1.0725	1.0833	1.0942
0.0	ϵ^{-x}	1.0000	0.9900	0.9802	0.9704	0.9608	0.9512	0.9418	0.9324	0.9231	0.9139
0.1	ϵ^x	1.1052	1.1163	1.1275	1.1388	1.1503	1.1618	1.1735	1.1853	1.1972	1.2093
0.1	ϵ^{-x}	0.9048	0.8958	0.8869	0.8781	0.8694	0.8607	0.8521	0.8437	0.8353	0.8270
0.2	ϵ^x	1.2214	1.2337	1.2461	1.2586	1.2712	1.2840	1.2969	1.3100	1.3231	1.3364
0.2	ϵ^{-x}	0.8187	0.8106	0.8025	0.7945	0.7866	0.7788	0.7711	0.7634	0.7558	0.7483
0.3	ϵ^x	1.3499	1.3634	1.3771	1.3910	1.4049	1.4191	1.4333	1.4477	1.4623	1.4770
0.3	ϵ^{-x}	0.7408	0.7334	0.7261	0.7189	0.7118	0.7047	0.6977	0.6907	0.6839	0.6771
0.4	ϵ^x	1.4918	1.5068	1.5220	1.5373	1.5527	1.5683	1.5841	1.6000	1.6161	1.6323
0.4	ϵ^{-x}	0.6703	0.6637	0.6570	0.6505	0.6440	0.6376	0.6313	0.6250	0.6188	0.6126
0.5	ϵ^x	1.6487	1.6653	1.6820	1.6989	1.7160	1.7333	1.7507	1.7683	1.7860	1.8040
0.5	ϵ^{-x}	0.6065	0.6005	0.5945	0.5886	0.5827	0.5769	0.5712	0.5655	0.5599	0.5543
0.6	ϵ^x	1.8221	1.8404	1.8589	1.8776	1.8965	1.9155	1.9348	1.9542	1.9739	1.9939
0.6	ϵ^{-x}	0.5488	0.5434	0.5379	0.5326	0.5273	0.5220	0.5169	0.5117	0.5066	0.5017
0.7	ϵ^x	2.0138	2.0340	2.0544	2.0751	2.0959	2.1170	2.1383	2.1598	2.1815	2.2034
0.7	ϵ^{-x}	0.4966	0.4916	0.4868	0.4819	0.4771	0.4724	0.4677	0.4630	0.4584	0.4538
0.8	ϵ^x	2.2255	2.2479	2.2705	2.2933	2.3164	2.3396	2.3632	2.3869	2.4109	2.4351
0.8	ϵ^{-x}	0.4493	0.4449	0.4404	0.4360	0.4317	0.4274	0.4232	0.4190	0.4148	0.4107
0.9	ϵ^x	2.4596	2.4843	2.5093	2.5345	2.5600	2.5857	2.6117	2.6379	2.6645	2.6912
0.9	ϵ^{-x}	0.4066	0.4025	0.3985	0.3946	0.3906	0.3867	0.3829	0.3791	0.3753	0.3716
1.0	ϵ^x	2.7183	2.7456	2.7732	2.8011	2.8292	2.8577	2.8864	2.9154	2.9447	2.9743
1.0	ϵ^{-x}	0.3679	0.3642	0.3606	0.3570	0.3535	0.3499	0.3465	0.3430	0.3396	0.3362
1.1	ϵ^x	3.0042	3.0344	3.0649	3.0957	3.1268	3.1582	3.1899	3.2220	3.2544	3.2871
1.1	ϵ^{-x}	0.3329	0.3296	0.3263	0.3230	0.3198	0.3166	0.3135	0.3104	0.3073	0.3042
1.2	ϵ^x	3.3201	3.3535	3.3872	3.4212	3.4556	3.4903	3.5254	3.5609	3.5966	3.6328
1.2	ϵ^{-x}	0.3012	0.2982	0.2952	0.2923	0.2894	0.2865	0.2837	0.2808	0.2780	0.2753
1.3	ϵ^x	3.6693	3.7062	3.7434	3.7810	3.8190	3.8574	3.8962	3.9354	3.9749	4.0149
1.3	ϵ^{-x}	0.2725	0.2698	0.2671	0.2645	0.2618	0.2592	0.2567	0.2541	0.2516	0.2491
1.4	ϵ^x	4.0552	4.0960	4.1371	4.1787	4.2207	4.2631	4.3060	4.3492	4.3929	4.4371
1.4	ϵ^{-x}	0.2466	0.2441	0.2417	0.2393	0.2369	0.2346	0.2322	0.2299	0.2276	0.2254
1.5	ϵ^x	4.4817	4.5267	4.5722	4.6182	4.6646	4.7115	4.7588	4.8066	4.8550	4.9037
1.5	ϵ^{-x}	0.2231	0.2209	0.2187	0.2165	0.2144	0.2122	0.2101	0.2080	0.2060	0.2039
1.6	ϵ^x	4.9530	5.0028	5.0531	5.1039	5.1552	5.2070	5.2593	5.3122	5.3656	5.4195
1.6	ϵ^{-x}	0.2019	0.1999	0.1979	0.1959	0.1940	0.1920	0.1901	0.1882	0.1864	0.1845
1.7	ϵ^x	5.4739	5.5290	5.5845	5.6407	5.6973	5.7546	5.8124	5.8709	5.9299	5.9895
1.7	ϵ^{-x}	0.1827	0.1809	0.1791	0.1773	0.1755	0.1738	0.1720	0.1703	0.1686	0.1670
1.8	ϵ^x	6.0496	6.1104	6.1719	6.2339	6.2965	6.3598	6.4237	6.4883	6.5535	6.6194
1.8	ϵ^{-x}	0.1653	0.1637	0.1620	0.1604	0.1588	0.1572	0.1557	0.1541	0.1526	0.1511
1.9	ϵ^x	6.6859	6.7531	6.8210	6.8895	6.9588	7.0287	7.0993	7.1707	7.2427	7.3155
1.9	ϵ^{-x}	0.1496	0.1481	0.1466	0.1451	0.1437	0.1423	0.1409	0.1395	0.1381	0.1367

Values of e^x and e^{-x}

x	Function	0.00	0.01	0.02	0.03	0.04	0.05	0.06	0.07	0.08	0.09
2.0	e^x	7.3891	7.4633	7.5383	7.6141	7.6906	7.7679	7.8460	7.9248	8.0045	8.0849
	e^{-x}	0.1353	0.1340	0.1327	0.1313	0.1300	0.1287	0.1275	0.1262	0.1249	0.1237
2.1	e^x	8.1662	8.2482	8.3311	8.4149	8.4994	8.5849	8.6711	8.7583	8.8463	8.9352
	e^{-x}	0.1225	0.1212	0.1200	0.1188	0.1177	0.1165	0.1153	0.1142	0.1130	0.1119
2.2	e^x	9.0250	9.1157	9.2073	9.2999	9.3933	9.4877	9.5831	9.6794	9.7767	9.8749
	e^{-x}	0.1108	0.1097	0.1086	0.1075	0.1065	0.1054	0.1044	0.1033	0.1023	0.1013
2.3	e^x	9.9742	10.074	10.176	10.278	10.381	10.486	10.591	10.697	10.805	10.913
	e^{-x}	0.1003	0.0993	0.0983	0.0973	0.0963	0.0954	0.0944	0.0935	0.0926	0.0916
2.4	e^x	11.023	11.134	11.246	11.359	11.473	11.588	11.705	11.822	11.941	12.061
	e^{-x}	0.0907	0.0898	0.0889	0.0880	0.0872	0.0863	0.0854	0.0846	0.0837	0.0829
2.5	e^x	12.182	12.305	12.429	12.554	12.680	12.807	12.936	13.066	13.197	13.330
	e^{-x}	0.0821	0.0813	0.0805	0.0797	0.0789	0.0781	0.0773	0.0765	0.0758	0.0750
2.6	e^x	13.464	13.599	13.736	13.874	14.013	14.154	14.296	14.440	14.585	14.732
	e^{-x}	0.0743	0.0735	0.0728	0.0721	0.0714	0.0707	0.0699	0.0693	0.0686	0.0679
2.7	e^x	14.880	15.029	15.180	15.333	15.487	15.643	15.800	15.959	16.119	16.281
	e^{-x}	0.0672	0.0665	0.0659	0.0652	0.0646	0.0639	0.0633	0.0627	0.0620	0.0614
2.8	e^x	16.445	16.610	16.777	16.945	17.116	17.288	17.462	17.637	17.814	17.993
	e^{-x}	0.0608	0.0602	0.0596	0.0590	0.0584	0.0578	0.0573	0.0567	0.0561	0.0556
2.9	e^x	18.174	18.357	18.541	18.728	18.916	19.106	19.298	19.492	19.688	19.886
	e^{-x}	0.0550	0.0545	0.0539	0.0534	0.0529	0.0523	0.0518	0.0513	0.0508	0.0503
3.0	e^x	20.086	20.287	20.491	20.697	20.905	21.115	21.328	21.542	21.758	21.977
	e^{-x}	0.0498	0.0493	0.0488	0.0483	0.0478	0.0474	0.0469	0.0464	0.0460	0.0455
3.1	e^x	22.198	22.421	22.646	22.874	23.104	23.336	23.571	23.807	24.047	24.288
	e^{-x}	0.0450	0.0446	0.0442	0.0437	0.0433	0.0429	0.0424	0.0420	0.0416	0.0412
3.2	e^x	24.533	24.779	25.028	25.280	25.534	25.790	26.050	26.311	26.576	26.843
	e^{-x}	0.0408	0.0404	0.0400	0.0396	0.0392	0.0388	0.0384	0.0380	0.0376	0.0373
3.3	e^x	27.113	27.385	27.660	27.938	28.219	28.503	28.789	29.079	29.371	29.666
	e^{-x}	0.0369	0.0365	0.0362	0.0358	0.0354	0.0351	0.0347	0.0344	0.0340	0.0337
3.4	e^x	29.964	30.265	30.569	30.877	31.187	31.500	31.817	32.137	32.460	32.786
	e^{-x}	0.0334	0.0330	0.0327	0.0324	0.0321	0.0317	0.0314	0.0311	0.0308	0.0305
3.5	e^x	33.115	33.448	33.784	34.124	34.467	34.813	35.163	35.517	35.874	36.234
	e^{-x}	0.0302	0.0299	0.0296	0.0293	0.0290	0.0287	0.0284	0.0282	0.0279	0.0276
3.6	e^x	36.598	36.966	37.338	37.713	38.092	38.475	38.861	39.252	39.646	40.045
	e^{-x}	0.0273	0.0271	0.0268	0.0265	0.0263	0.0260	0.0257	0.0255	0.0252	0.0250
3.7	e^x	40.447	40.854	41.264	41.679	42.098	42.521	42.948	43.380	43.816	44.256
	e^{-x}	0.0247	0.0245	0.0242	0.0240	0.0238	0.0235	0.0233	0.0231	0.0228	0.0226
3.8	e^x	44.701	45.150	45.604	46.063	46.525	46.993	47.465	47.942	48.424	48.911
	e^{-x}	0.0224	0.0221	0.0219	0.0217	0.0215	0.0213	0.0211	0.0209	0.0207	0.0204
3.9	e^x	49.402	49.899	50.400	50.907	51.419	51.935	52.457	52.985	53.517	54.055
	e^{-x}	0.0202	0.0200	0.0198	0.0196	0.0195	0.0193	0.0191	0.0189	0.0187	0.0185

Values of e^x and e^{-x}

x	Function	0.00	0.01	0.02	0.03	0.04	0.05	0.06	0.07	0.08	0.09
4.0	e^x	54.598	55.147	55.701	56.261	56.826	57.397	57.974	58.557	59.145	59.740
	e^{-x}	0.0183	0.0181	0.0180	0.0178	0.0176	0.0174	0.0172	0.0171	0.0169	0.0167
4.1	e^x	60.340	60.947	61.559	62.178	62.803	63.434	64.072	64.715	65.366	66.023
	e^{-x}	0.0166	0.0164	0.0162	0.0161	0.0159	0.0158	0.0156	0.0155	0.0153	0.0151
4.2	e^x	66.686	67.357	68.033	68.717	69.408	70.105	70.810	71.522	72.240	72.966
	e^{-x}	0.0150	0.0148	0.0147	0.0146	0.0144	0.0143	0.0141	0.0140	0.0138	0.0137
4.3	e^x	73.700	74.440	75.189	75.944	76.708	77.478	78.257	79.044	79.838	80.640
	e^{-x}	0.0136	0.0134	0.0133	0.0132	0.0130	0.0129	0.0128	0.0127	0.0125	0.0124
4.4	e^x	81.451	82.269	83.096	83.931	84.775	85.627	86.488	87.357	88.235	89.121
	e^{-x}	0.0123	0.0122	0.0120	0.0119	0.0118	0.0117	0.0116	0.0114	0.0113	0.0112
4.5	e^x	90.017	90.922	91.836	92.759	93.691	94.632	95.583	96.544	97.514	98.494
	e^{-x}	0.0111	0.0110	0.0109	0.0108	0.0107	0.0106	0.0105	0.0104	0.0103	0.0102
4.6	e^x	99.484	100.48	101.49	102.51	103.54	104.58	105.64	106.70	107.77	108.85
	e^{-x}	0.0101	0.0100	0.0099	0.0098	0.0097	0.0096	0.0095	0.0094	0.0093	0.0092
4.7	e^x	109.95	111.05	112.17	113.30	114.43	115.58	116.75	117.92	119.10	120.30
	e^{-x}	0.0091	0.0090	0.0089	0.0088	0.0087	0.0087	0.0086	0.0085	0.0084	0.0083
4.8	e^x	121.51	122.73	123.97	125.21	126.47	127.74	129.02	130.32	131.63	132.95
	e^{-x}	0.0082	0.0081	0.0081	0.0080	0.0079	0.0078	0.0078	0.0077	0.0076	0.0075
4.9	e^x	134.29	135.64	137.00	138.38	139.77	141.17	142.59	144.03	145.47	146.94
	e^{-x}	0.0074	0.0074	0.0073	0.0072	0.0072	0.0071	0.0070	0.0069	0.0069	0.0068
5.0	e^x	148.41	149.90	151.41	152.93	154.47	156.02	157.59	159.17	160.77	162.39
	e^{-x}	0.0067	0.0067	0.0066	0.0065	0.0065	0.0064	0.0063	0.0063	0.0062	0.0062
5.1	e^x	164.02	165.67	167.34	169.02	170.72	172.43	174.16	175.91	177.68	179.47
	e^{-x}	0.0061	0.0060	0.0060	0.0059	0.0059	0.0058	0.0057	0.0057	0.0056	0.0056
5.2	e^x	181.27	183.09	184.93	186.79	188.67	190.57	192.48	194.42	196.37	198.34
	e^{-x}	0.0055	0.0055	0.0054	0.0054	0.0053	0.0052	0.0052	0.0051	0.0051	0.0050
5.3	e^x	200.34	202.35	204.38	206.44	208.51	210.61	212.72	214.86	217.02	219.20
	e^{-x}	0.0050	0.0049	0.0049	0.0048	0.0048	0.0047	0.0047	0.0047	0.0046	0.0046
5.4	e^x	221.41	223.63	225.88	228.15	230.44	232.76	235.10	237.46	239.85	242.26
	e^{-x}	0.0045	0.0045	0.0044	0.0044	0.0043	0.0043	0.0043	0.0042	0.0042	0.0041
5.5	e^x	244.69	247.15	249.64	252.14	254.68	257.24	259.82	262.43	265.07	267.74
	e^{-x}	0.0041	0.0040	0.0040	0.0040	0.0039	0.0039	0.0038	0.0038	0.0038	0.0037
5.6	e^x	270.43	273.14	275.89	278.66	281.46	284.29	287.15	290.03	292.95	295.89
	e^{-x}	0.0037	0.0037	0.0036	0.0036	0.0036	0.0035	0.0035	0.0034	0.0034	0.0034
5.7	e^x	298.87	301.87	304.90	307.97	311.06	314.19	317.35	320.54	323.76	327.01
	e^{-x}	0.0033	0.0033	0.0033	0.0032	0.0032	0.0032	0.0032	0.0031	0.0031	0.0031
5.8	e^x	330.30	333.62	336.97	340.36	343.78	347.23	350.72	354.25	357.81	361.41
	e^{-x}	0.0030	0.0030	0.0030	0.0029	0.0029	0.0029	0.0029	0.0028	0.0028	0.0028
5.9	e^x	365.04	368.71	372.41	376.15	379.93	383.75	387.61	391.51	395.44	399.41
	e^{-x}	0.0027	0.0027	0.0027	0.0027	0.0026	0.0026	0.0026	0.0026	0.0025	0.0025

Appendix E
Trigonometric Functions

Degrees	Sine	Tangent	Cotangent	Cosine	
0	.0000	.0000	......	1.0000	90
1	.0175	.0175	57.29	.9998	89
2	.0349	.0349	28.636	.9994	88
3	.0523	.0524	19.081	.9986	87
4	.0698	.0699	14.301	.9976	86
5	.0872	.0875	11.430	.9962	85
6	.1045	.1051	9.5144	.9945	84
7	.1219	.1228	8.1443	.9925	83
8	.1392	.1405	7.1154	.9903	82
9	.1564	.1584	6.3138	.9877	81
10	.1736	.1763	5.6713	.9848	80
11	.1908	.1944	5.1446	.9816	79
12	.2079	.2126	4.7046	.9781	78
13	.2250	.2309	4.3315	.9744	77
14	.2419	.2493	4.0108	.9703	76
15	.2588	.2679	3.7321	.9659	75
16	.2756	.2867	3.4874	.9613	74
17	.2924	.3057	3.2709	.9563	73
18	.3090	.3249	3.0777	.9511	72
19	.3256	.3443	2.9042	.9455	71
20	.3420	.3640	2.7475	.9397	70
21	.3584	.3839	2.6051	.9336	69
22	.3746	.4040	2.4751	.9272	68
23	.3907	.4245	2.3559	.9205	67
24	.4067	.4452	2.2460	.9135	66
25	.4226	.4663	2.1445	.9063	65
26	.4384	.4877	2.0503	.8988	64
27	.4540	.5095	1.9626	.8910	63
28	.4695	.5317	1.8807	.8829	62
29	.4848	.5543	1.8040	.8746	61
30	.5000	.5774	1.7321	.8660	60
31	.5150	.6009	1.6643	.8572	59
32	.5299	.6249	1.6003	.8480	58
33	.5446	.6494	1.5399	.8387	57
34	.5592	.6745	1.4826	.8290	56
35	.5736	.7002	1.4281	.8192	55
36	.5878	.7265	1.3764	.8090	54
37	.6018	.7536	1.3270	.7986	53
38	.6157	.7813	1.2799	.7880	52
39	.6293	.8098	1.2349	.7771	51
40	.6428	.8391	1.1918	.7660	50
41	.6561	.8693	1.1504	.7547	49
42	.6691	.9004	1.1106	.7431	48
43	.6820	.9325	1.0724	.7314	47
44	.6947	.9657	1.0355	.7193	46
45	.7071	1.0000	1.0000	.7071	45
	Cosine	Cotangent	Tangent	Sine	Degrees

Appendix F
Powers of Numbers

n	n^4	n^5	n^6	n^7	n^8
1	1	1	1	1	1
2	16	32	64	128	256
3	81	243	729	2187	6561
4	256	1024	4096	16384	65536
5	625	3125	15625	78125	390625
6	1296	7776	46656	279936	1679616
7	2401	16807	117649	823543	5764801
8	4096	32768	262144	2097152	16777216
9	6561	59049	531441	4782969	43046721
					$\times 10^8$
10	10000	100000	1000000	10000000	1.000000
11	14641	161051	1771561	19487171	2.143589
12	20736	248832	2985984	35831808	4.299817
13	28561	371293	4826809	62748517	8.157307
14	38416	537824	7529536	105413504	14.757891
15	50625	759375	11390625	170859375	25.628906
16	65536	1048576	16777216	268435456	42.949673
17	83521	1419857	24137569	410338673	69.757574
18	104976	1889568	34012224	612220032	110.199606
19	130321	2476099	47045881	893871739	169.835630
				$\times 10^9$	$\times 10^{10}$
20	160000	3200000	64000000	1.280000	2.560000
21	194481	4084101	85766121	1.801089	3.782286
22	234256	5153632	113379904	2.494358	5.487587
23	279841	6436343	148035889	3.404825	7.831099
24	331776	7962624	191102976	4.586471	11.007531
25	390625	9765625	244140625	6.103516	15.258789
26	456976	11881376	308915776	8.031810	20.882706
27	531441	14348907	387420489	10.460353	28.242954
28	614656	17210368	481890304	13.492929	37.780200
29	707281	20511149	594823321	17.249876	50.024641
			$\times 10^8$	$\times 10^{10}$	$\times 10^{11}$
30	810000	24300000	7.290000	2.187000	6.561000
31	923521	28629151	8.875037	2.751261	8.528910
32	1048576	33554432	10.737418	3.435974	10.995116
33	1185921	39135393	12.914680	4.261844	14.064086
34	1336336	45435424	15.448044	5.252335	17.857939
35	1500625	52521875	18.382656	6.433930	22.518754
36	1679616	60466176	21.767823	7.836416	28.211099
37	1874161	69343957	25.657264	9.493188	35.124795
38	2085136	79235168	30.109364	11.441558	43.477921
39	2313441	90224199	35.187438	13.723101	53.520093
			$\times 10^9$	$\times 10^{10}$	$\times 10^{12}$
40	2560000	102400000	4.096000	16.384000	6.553600
41	2825761	115856201	4.750104	19.475427	7.984925
42	3111696	130691232	5.489032	23.053933	9.682652
43	3418801	147008443	6.321363	27.181861	11.688200
44	3748096	164916224	7.256314	31.927781	14.048224
45	4100625	184528125	8.303766	37.366945	16.815125
46	4477456	205962976	9.474297	43.581766	20.047612
47	4879681	229345007	10.779215	50.662312	23.811287
48	5308416	254803968	12.230590	58.706834	28.179280
49	5764801	282475249	13.841287	67.822307	33.232931
50	6250000	312500000	15.625000	78.125000	39.062500

POWERS OF NUMBERS—Continued

n	n^4	n^5	n^6	n^7	n^8
			$\times 10^9$	$\times 10^{11}$	$\times 10^{13}$
50	6250000	312500000	15.625000	7.812500	3.906250
51	6765201	345025251	17.596288	8.974107	4.576794
52	7311616	380204032	19.770610	10.280717	5.345973
53	7890481	418195493	22.164361	11.747111	6.225969
54	8503056	459165024	24.794911	13.389252	7.230196
55	9150625	503284375	27.680641	15.224352	8.373394
56	9834496	550731776	30.840979	17.270948	9.671731
57	10556001	601692057	34.296447	19.548975	11.142916
58	11316496	656356768	38.068693	22.079842	12.806308
59	12117361	714924299	42.180534	24.886515	14.683044
		$\times 10^8$	$\times 10^{10}$	$\times 10^{11}$	$\times 10^{13}$
60	12960000	7.776000	4.665600	27.993600	16.796160
61	13845841	8.445963	5.152037	31.427428	19.170731
62	14776336	9.161328	5.680024	35.216146	21.834011
63	15752961	9.924365	6.252350	39.389806	24.815578
64	16777216	10.737418	6.871948	43.980465	28.147498
65	17850625	11.602906	7.541889	49.022279	31.864481
66	18974736	12.523326	8.265395	54.551607	36.004061
67	20151121	13.501251	9.045838	60.607116	40.606768
68	21381376	14.539336	9.886748	67.229888	45.716324
69	22667121	15.640313	10.791816	74.463533	51.379837
		$\times 10^8$	$\times 10^{10}$	$\times 10^{12}$	$\times 10^{14}$
70	24010000	16.807000	11.764900	8.235430	5.764801
71	25411681	18.042294	12.810028	9.095120	6.457535
72	26873856	19.349176	13.931407	10.030613	7.222041
73	28398241	20.730716	15.133423	11.047399	8.064601
74	29986576	22.190066	16.420649	12.151280	8.991947
75	31640625	23.730469	17.797852	13.348389	10.011292
76	33362176	25.355254	19.269993	14.645195	11.130348
77	35153041	27.067842	20.842238	16.048523	12.357363
78	37015056	28.871744	22.519960	17.565569	13.701144
79	38950081	30.770564	24.308746	19.203909	15.171088
		$\times 10^8$	$\times 10^{10}$	$\times 10^{12}$	$\times 10^{14}$
80	40960000	32.768000	26.214400	20.971520	16.777216
81	43046721	34.867844	28.242954	22.876792	18.530202
82	45212176	37.073984	30.400667	24.928547	20.441409
83	47458321	39.390406	32.694037	27.136051	22.522922
84	49787136	41.821194	35.129803	29.509035	24.787589
85	52200625	44.370531	37.714952	32.057709	27.249053
86	54700816	47.042702	40.456724	34.792782	29.921793
87	57289761	49.842092	43.362620	37.725479	32.821167
88	59969536	52.773192	46.440409	40.867560	35.963452
89	62742241	55.840594	49.698129	44.231335	39.365888
		$\times 10^9$	$\times 10^{11}$	$\times 10^{13}$	$\times 10^{15}$
90	65610000	5.904900	5.314410	4.782969	4.304672
91	68574961	6.240321	5.678693	5.167610	4.702525
92	71639296	6.590815	6.063550	5.578466	5.132189
93	74805201	6.956884	6.469902	6.017009	5.595818
94	78074896	7.339040	6.898698	6.484776	6.095689
95	81450625	7.737809	7.350919	6.983373	6.634204
96	84934656	8.153727	7.827578	7.514475	7.213896
97	88529281	8.587340	8.329720	8.079828	7.837434
98	92236816	9.039208	8.858424	8.681255	8.507630
99	96059601	9.509900	9.414801	9.320653	9.227447
100	100000000	10.000000	10.000000	10.000000	10.000000

Index

A

A priori probability	166
Acceleration	222
Accuracy and precision	75
Addition	
operator	104
tables	102
Advanced numeric math	184
Aleph	207
Algebra and graphic art	211
Ancestor	80
Antilogs	122
Arithmetic verbs	102
Arrays	34
Averaging	186

B

Banking model	141
Bases	54
conversions	54
3 conversion	56
Binary conversion	55
Boolean algebra	194
Borrow	107

C

Calculus of	
algebra	210
arithmetic	209
Cardinal numbers	62
Carry digit	98
Change model	140
Commutativity	45
Complements	34
Confidence level	93
Continuous numbers	67
Convergent series	134, 216
Cryptograms	43

D

Decimal numbers	53
Definite integrals	238
Definition	29
by enumeration	31
Delta process	224
Denumerably infinite sets	205
Dependent variables	100
Derivatives	225
Difference	107
Differential calculus applications	227
Differentiation	224
Digits	12
Divergent series	216
Division tables	109

E

Empirical probability	166
Equality	86

Equations	94
construction	95
systems	97
Euler's equation	199
Exponent	115

F

Factorials	126
Folium of DesCartes	215
Force/direction model	148
Formal logic	22
Functions	81

G

Geometric mean	190
Greater than	86

H

Harmonic mean	188
History of math	8
Hollerith cards	18
Hyperbola	214

I

Imaginary numbers	72
Indefinite integrals	237
Independent variables	100
Induction	126
Infinite sets	202
Infinity	51
Integral calculus	236
Irrational numbers	65
Iteration	125

L

Law of	
absorption	196
thought	194
Least significant digit	61
Less than	86
Logarithms	119
Logic	167
Logical	
product	195
sum	195

M

Mappings	45, 167
Math language	21

Mathematical	
models	136
truth	89
Mean	186
Median	188
Mileage model	145
Minuend	107
Moebius strip	200
Most significant digit	60
Multiplication tables	108

N

Naming	96
Negative numbers	64
Nonnumeric mathematics	193
Numbers	12
definition	49
line	69, 181
names	52
puzzes	179
Numerals	12
definition	49

O

Octal conversion	57
Open interval	219
Operands	26
Operation selection	137
Operators	27
Order	42

P

Parabola	213
Paradox	170
Partial ordering	88
Pathological functions	225
Pattern	
matching	147
rotation	129
Perimeter/area calculations	154
Pi	111
Positional notation	59, 103
Powers	113
Predecessor	80
Probability	161, 190
Problem description	137
Product set	44
Progressions	132
Proper subset	32
Properties of relations	82
Puzzles	167

Q

Quotient	111

R

Radix	54
Rate of change	220
Rational numbers	62
Ratios	64
Reciprocals	133
Reductio ad absurdum	92
Reflexivity	85
Relations	42, 78
Remainder	110
Rigor	91
Root	115
extraction	115
Roundoff error	123

S

Scatter	187
Sequence of summation	135
Sets	28
characteristics	31
classification	173
diagrams	38
division	40
intersections	36
operators	34
theory	28
theory tools	38
union	37
Signed intergers	64
Simultaneous equations	100
Statistics	185
Subsets	31
Subtraction tables	107
Subtrahend	107
Successors	79
function	50
Supersets	34
Symbolic logic	196
Symmetry	84

T

Tautology	90
Topology	198
Transcendental numbers	68
Transfinite math	202
Transivity	84

U

Union	195
Universe	33, 39

V

Variable of induction	126
Velocity	222
Venn diagrams	39